Weather

629.130684 LANKFORD
Lankford, Terry T.
Weather /

JUL 27 2002

RIDGLEA BRANCH

The McGraw-Hill *CONTROLLING PILOT ERROR* Series

Weather
Terry T. Lankford

Communications
Paul E. Illman

Automation
Vladimir Risukhin

Controlled Flight into Terrain (CFIT/CFTT)
Daryl R. Smith

Training and Instruction
David A. Frazier

Checklists and Compliance
Thomas P. Turner

Maintenance and Mechanics
Larry Reithmaier

Situational Awareness
Paul A. Craig

Fatigue
James C. Miller

Culture, Environment, and CRM
Tony Kern

Cover Photo Credits (clockwise from upper left): PhotoDisc; Corbis Images; from *Spin Management and Recovery* by Michael C. Love; PhotoDisc; PhotoDisc; PhotoDisc; image by Kelly Parr; © 2001 Mike Fizer, all rights reserved; *Piloting Basics Handbook* by Bjork, courtesy of McGraw-Hill; PhotoDisc.

CONTROLLING PILOT ERROR

Weather

Terry T. Lankford

McGraw-Hill

New York   Chicago   San Francisco   Lisbon   London   Madrid
Mexico City   Milan   New Delhi   San Juan   Seoul
Singapore   Sydney   Toronto

Cataloging-in-Publication Data is on file with the Library of Congress

McGraw-Hill

A Division of The McGraw·Hill Companies

Copyright © 2001 by The McGraw-Hill Companies, Inc. All rights reserved. Printed in the United States of America. Except as permitted under the United States Copyright Act of 1976, no part of this publication may be reproduced or distributed in any form or by any means, or stored in a data base or retrieval system, without the prior written permission of the publisher.

1 2 3 4 5 6 7 8 9 0   DOC/DOC   0 9 8 7 6 5 4 3 2 1

ISBN 0-07-137328-4

The sponsoring editor for this book was Shelley Ingram Carr, the editing supervisor was Steven Melvin, and the production supervisor was Pamela Pelton. It was set in Garamond per the TAB3A design by Joanne Morbit of McGraw-Hill's Hightstown, N.J., Professional Book Group composition unit.

Printed and bound by R. R. Donnelley & Sons Company.

 This book is printed on recycled, acid-free paper containing a minimum of 50% recycled de-inked fiber.

McGraw-Hill books are available at special quantity discounts to use as premiums and sales promotions, or for use in corporate training programs. For more information, please write to the Director of Special Sales, Professional Publishing, McGraw-Hill, Two Penn Plaza, New York, NY 10121-2298. Or contact your local bookstore.

Information contained in this work has been obtained by The McGraw-Hill Companies, Inc. ("McGraw-Hill") from sources believed to be reliable. However, neither McGraw-Hill nor its authors guarantee the accuracy or completeness of any information published herein and neither McGraw-Hill nor its authors shall be responsible for any errors, omissions, or damages arising out of use of this information. This work is published with the understanding that McGraw-Hill and its authors are supplying information but are not attempting to render engineering or other professional services. If such services are required, the assistance of an appropriate professional should be sought.

Contents

Series Introduction *vii*

Preface *xvii*

1 **Introduction** *1*
2 **Automated Weather Observations** *9*
3 **Low Ceilings and Visibility** *25*
4 **Turbulence** *43*
5 **Icing** *63*
6 **Thunderstorms** *85*
7 **Weather Systems** *101*
8 **Aircraft Performance** *117*
9 **Summary** *131*
   Index *148*

Series Introduction

The Human Condition

The Roman philosopher Cicero may have been the first to record the much-quoted phrase "to err is human." Since that time, for nearly 2000 years, the malady of human error has played out in triumph and tragedy. It has been the subject of countless doctoral dissertations, books, and, more recently, television documentaries such as "History's Greatest Military Blunders." Aviation is not exempt from this scrutiny, as evidenced by the excellent Learning Channel documentary "Blame the Pilot" or the NOVA special "Why Planes Crash," featuring John Nance. Indeed, error is so prevalent throughout history that our flaws have become associated with our very being, hence the phrase *the human condition*.

The Purpose of This Series

Simply stated, the purpose of the Controlling Pilot Error series is to address the so-called human condition, improve performance in aviation, and, in so doing, save a few lives. It is not our intent to rehash the work of over a millennia of expert and amateur opinions but rather to *apply* some of the more important and insightful theoretical perspectives to the life and death arena of manned flight. To the best of my knowledge, no effort of this magnitude has ever been attempted in aviation, or anywhere else for that matter. What follows is an extraordinary combination of why, what, and how to avoid and control error in aviation.

Because most pilots are practical people at heart—many of whom like to spin a yarn over a cold lager—we will apply this wisdom to the daily flight environment, using a case study approach. The vast majority of the case studies you will read are taken directly from aviators who have made mistakes (or have been victimized by the mistakes of others) and survived to tell about it. Further to their credit, they have reported these events via the anonymous Aviation Safety Reporting System (ASRS), an outstanding program that provides a wealth of extremely useful and *usable* data to those who seek to make the skies a safer place.

A Brief Word about the ASRS

The ASRS was established in 1975 under a Memorandum of Agreement between the Federal Aviation Administration (FAA) and the National Aeronautics and Space Administration (NASA). According to the official ASRS web site, *http://asrs.arc.nasa.gov*

> The ASRS collects, analyzes, and responds to voluntarily submitted aviation safety incident reports in order to lessen the likelihood of aviation accidents. ASRS data are used to:
>
> - Identify deficiencies and discrepancies in the National Aviation System (NAS) so that these can be remedied by appropriate authorities.

- Support policy formulation and planning for, and improvements to, the NAS.
- Strengthen the foundation of aviation human factors safety research. This is particularly important since it is generally conceded *that over two-thirds of all aviation accidents and incidents have their roots in human performance errors* (emphasis added).

Certain types of analyses have already been done to the ASRS data to produce "data sets," or prepackaged groups of reports that have been screened "for the relevance to the topic description" (ASRS web site). These data sets serve as the foundation of our Controlling Pilot Error project. The data come *from* practitioners and are *for* practitioners.

The Great Debate

The title for this series was selected after much discussion and considerable debate. This is because many aviation professionals disagree about what should be done about the problem of pilot error. The debate is basically three sided. On one side are those who say we should seek any and all available means to *eliminate* human error from the cockpit. This effort takes on two forms. The first approach, backed by considerable capitalistic enthusiasm, is to automate human error out of the system. Literally billions of dollars are spent on so-called human-aiding technologies, high-tech systems such as the Ground Proximity Warning System (GPWS) and the Traffic Alert and Collision Avoidance System (TCAS). Although these systems have undoubtedly made the skies safer, some argue that they have made the pilot more complacent and dependent on the automation, creating an entirely new set of pilot errors. Already the automation enthusiasts are seeking robotic answers for this new challenge. Not surprisingly, many pilot trainers see the problem from a slightly different angle.

Another branch on the "eliminate error" side of the debate argues for higher training and education standards, more accountability, and better screening. This group (of which I count myself a member) argues that some industries (but not yet ours) simply don't make serious errors, or at least the errors are so infrequent that they are statistically nonexistent. This group asks, "How many errors should we allow those who handle nuclear weapons or highly dangerous viruses like Ebola or anthrax?" The group cites research on high-reliability organizations (HROs) and believes that aviation needs to be molded into the HRO mentality. (For more on high-reliability organizations, see "Culture, Environment, and CRM" in this series.) As you might expect, many status quo aviators don't warm quickly to these ideas for more education, training, and accountability—and point to their excellent safety records to say such efforts are not needed. They recommend a different approach, one where no one is really at fault.

On the far opposite side of the debate lie those who argue for "blameless cultures" and "error-tolerant systems." This group agrees with Cicero that "to err is human" and advocates "error-management," a concept that prepares pilots to recognize and "trap" error before it can build upon itself into a mishap chain of events. The group feels that training should be focused on primarily error mitigation rather than (or, in some cases, in addition to) error prevention.

Falling somewhere between these two extremes are two less-radical but still opposing ideas. The first approach is designed to prevent a recurring error. It goes something like this: "Pilot X did this or that and it led to a mishap, so don't do what Pilot X did." Regulators are particularly fond of this approach, and they attempt to regulate the last mishap out of future existence. These so-called rules written in blood provide the traditionalist with plenty of training materials and even come with ready-made case studies—the mishap that precipitated the rule.

Opponents to this "last mishap" philosophy argue for a more positive approach, one where we educate and train *toward* a complete set of known and valid competencies (positive behaviors) instead of seeking to eliminate negative behaviors. This group argues that the professional airmanship potential of the vast majority of our aviators is seldom approached—let alone realized. This was the subject of an earlier McGraw-Hill release, *Redefining Airmanship*.[1]

Who's Right? Who's Wrong? Who Cares?

It's not about *who's* right, but rather *what's* right. Taking the philosophy that there is value in all sides of a debate, the Controlling Pilot Error series is the first truly comprehensive approach to pilot error. By taking a unique "before-during-after" approach and using modern-era case studies, 10 authors—each an expert in the subject at hand—methodically attack the problem of pilot error from several angles. First, they focus on error prevention by taking a case study and showing how preemptive education and training, applied to planning and execution, could have avoided the error entirely. Second, the authors apply error management principles to the case study to show how a mistake could have been (or was) mitigated after it was made. Finally, the case study participants are treated to a thorough "debrief," where alternatives are discussed to prevent a reoccurrence of the error. By analyzing the conditions before, during, and after each case study, we hope to combine the best of all areas of the error-prevention debate.

A Word on Authors and Format

Topics and authors for this series were carefully analyzed and hand-picked. As mentioned earlier, the topics were taken from preculled data sets and selected for their relevance by NASA-Ames scientists. The authors were chosen for their interest and expertise in the given topic area. Some are experienced authors and researchers, but, more important, *all* are highly experienced in the aviation field about which they are writing. In a word, they are practitioners and have "been there and done that" as it relates to their particular topic.

In many cases, the authors have chosen to expand on the ASRS reports with case studies from a variety of sources, including their own experience. Although Controlling Pilot Error is designed as a comprehensive series, the reader should not expect complete uniformity of format or analytical approach. Each author has brought his own unique style and strengths to bear on the problem at hand. For this reason, each volume in the series can be used as a stand-alone reference or as a part of a complete library of common pilot error materials.

Although there are nearly as many ways to view pilot error as there are to make them, all authors were familiarized with what I personally believe should be the industry standard for the analysis of human error in aviation. The Human Factors Analysis and Classification System (HFACS) builds upon the groundbreaking and seminal work of James Reason to identify and organize human error into distinct and extremely useful subcategories. Scott Shappell and Doug Wiegmann completed the picture of error and error resistance by identifying common fail points in organizations and individuals. The following overview of this outstanding guide[2] to understanding pilot error is adapted from a United States Navy mishap investigation presentation.

> Simply writing off aviation mishaps to "aircrew error" is a simplistic, if not naive, approach to mishap causation. After all, it is well established that mishaps cannot be attributed to a single cause, or in most instances, even a single individual. Rather, accidents are the end result of a myriad of latent and active failures, only the last of which are the unsafe acts of the aircrew.
>
> As described by Reason,[3] active failures are the actions or inactions of operators that are believed to cause the accident. Traditionally referred to as "pilot error," they are the last "unsafe acts" committed by aircrew, often with immediate and tragic consequences. For example, forgetting to lower the landing gear before touch down or hotdogging through a box canyon will yield relatively immediate, and potentially grave, consequences.
>
> In contrast, latent failures are errors committed by individuals within the supervisory chain of command that effect the tragic sequence of events characteristic of an accident. For example, it is not difficult to understand how tasking aviators at the expense of quality crew rest can lead to fatigue and ultimately errors (active failures) in the cockpit. Viewed from this perspective then, the unsafe acts of aircrew are the end result of a long chain of causes whose roots originate in other parts (often the upper echelons) of the organization.

The problem is that these latent failures may lie dormant or undetected for hours, days, weeks, or longer until one day they bite the unsuspecting aircrew....

What makes the [Reason's] "Swiss Cheese" model particularly useful in any investigation of pilot error is that it forces investigators to address latent failures within the causal sequence of events as well. For instance, latent failures such as fatigue, complacency, illness, and the loss of

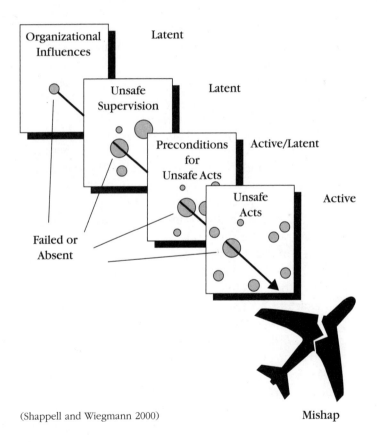

(Shappell and Wiegmann 2000)

situational awareness all effect performance but can be overlooked by investigators with even the best of intentions. These particular latent failures are described within the context of the "Swiss Cheese" model as preconditions for unsafe acts. Likewise, unsafe supervisory practices can promote unsafe conditions within operators and ultimately unsafe acts will occur. Regardless, whenever a mishap does occur, the crew naturally bears a great deal of the responsibility and must be held accountable. However, in many instances, the latent failures at the supervisory level were equally, if not more, responsible for the mishap. In a sense, the crew was set up for failure. . . .

But the "Swiss Cheese" model doesn't stop at the supervisory levels either, the organization itself can impact performance at all levels. For instance, in times of fiscal austerity funding is often cut, and as a result, training and flight time is curtailed. Supervisors are therefore left with tasking "non-proficient" aviators with sometimes-complex missions. Not surprisingly, causal factors such as task saturation and the loss of situational awareness will begin to appear and consequently performance in the cockpit will suffer. As such, causal factors at all levels must be addressed if any mishap investigation and prevention system is going to work.[4]

The HFACS serves as a reference for error interpretation throughout this series, and we gratefully acknowledge the works of Drs. Reason, Shappell, and Wiegmann in this effort.

No Time to Lose

So let us begin a journey together toward greater knowledge, improved awareness, and safer skies. Pick up any volume in this series and begin the process of self-analysis that is required for significant personal or organizational change. The complexity of the aviation environment demands a foundation of solid airmanship and a healthy, positive approach to combating pilot error. We believe this series will help you on this quest.

References

1. Kern, Tony, *Redefining Airmanship,* McGraw-Hill, New York, 1997.

2. Shappell, S. A., and Wiegmann, D. A., *The Human Factors Analysis and Classification System—HFACS,* DOT/FAA/AM-00/7, February 2000.

3. Reason, J. T., *Human Error,* Cambridge University Press, Cambridge, England, 1990.

4. U.S. Navy, *A Human Error Approach to Accident Investigation,* OPNAV 3750.6R, Appendix O, 2000.

Tony Kern

Foreword

It seems that no matter where I live, when the discussion turns to the local weather patterns, some local sage will announce "That's what's so great about [insert your city here], if you don't like the weather, wait 15 minutes and it will change." Everyone nods in agreement and I always laugh politely. Then I think to myself, *"If they only knew . . ."*

In aviation, weather changes occur in a matter of *seconds*, often with frightening, and sometimes lethal, implications. It's not that the information is unavailable. There are plenty of resources for obtaining weather information prior to and during flight. Meteorological and aeronautical information is provided by an alphabet soup of services such as the Transcribed Weather Broadcasts (TWEB), the Pilot's Automatic Telephone Answering Service (PATWAS), the Telephone Information Briefing Service (TIBS), in-flight advisories such as SIGMETs, Convective SIGMETs, Center Weather Advisories (CWA), Alert Weather Watches (AWW), and AIRMETs, and the Hazardous In-flight Weather Advisory Service (HIWAS) weather broadcasts routinely. There are also plenty of private services that provide excellent and up-to-date weather data. No, it is not typically a lack of information that gets a pilot in trouble, it seems to be a matter of convincing pilots to access and act prudently upon the information that is there at his or her fingertips.

I believe that even the term "aviation weather" is easily misconstrued. It conjures up pictures of towering cumulus clouds and intense microbursts. But the more prevalent risks may be more subtle; an increasing headwind that threatens an adequate fuel reserve, a sloping cloud deck that leads to disorientation, a downdraft on the lee side of a mountain pass.

Indeed, the most hazardous weather phenomena may seem mundane, even routine, like the unforecast, slowly developing ocean haze on a moonless night off the coast of Martha's Vineyard that would claim the life of America's favorite son, John F. Kennedy, Jr.

According to the NTSB the weather forecast that Kennedy got from the Internet about 2 hours before his flight offered no warning of the haze that hung over his route from New Jersey to Martha's Vineyard. The forecast was for good visual flying conditions with visibility of 6 to 8 miles. Yet, for a pilot with slightly more than 300 hours flight time, the combination of the moonless night and haze over the featureless ocean proved to be overwhelming. I wonder when he first suspected that he might be getting in over his head? What could he have done at that moment to change the outcome? Perhaps more to the point of this book, what can we do together to ensure that this doesn't happen to any of us?

In the pages that follow, you will come face to face with dozens of aviators who were more fortunate. They fooled around with Mother Nature and soon wished that they hadn't, but through good judgment, good hands, good fortune (or likely the combination of all three), these aviators lived to tell the tale, and we are the better for it.

We are led on this journey by one of the world's leading experts on the practical application of weather theory to day-to-day flight operations. Terry Lankford is a certified flight instructor and flight service station specialist, and he has written several books on the subject of weather interpretation and piloting, including *Weather Reports, Forecasts, and Flight Planning*; *Cockpit Weather Decisions*; and the *Aviation Weather Handbook*.

In this book, he carefully dissects the dynamic relationship between man, machine, and the environment, utilizing real-world examples to put you and me in the cockpit at the point of attack, learning from the experiences of others at groundspeed zero and in the safety of our favorite reading chair. Please use this opportunity to examine your own biases and habit patterns as they relate to weather flying. If we head into the sky unprepared, Mother Nature usually gets the last laugh.

Tony Kern

Preface

I obtained a Private Pilot Certificate in 1967 through an Air Force aero club in England—a certificate because the Federal Aviation Administration (FAA) can't spell license. Back in the States, with the G.I. Bill, I obtained commercial and flight instructor certificates, along with an instrument rating. Subsequently, as a full-time flight instructor I earned a Gold Seal. I have owned two airplanes, both Cessna 150s that have taken me across the country twice. I have also flown in Canada and Mexico.

A commercial pilot certificate and instrument rating qualified me for a position with the FAA as an Air Traffic Control Specialist (Station). This brought together my two interests: aviation and weather.

Only once as flight service specialist did I become an "air traffic controller." While at the Lovelock, Nevada FSS a call came in from a VFR pilot caught in clouds at 13,000 ft. In such cases the control facility (approach control or center) usually provides assistance. Coordinating with Oakland Center, the controller replied: "OK, you've got 14 and below, keep me advised." Wow, instant air traffic controller!

I retired from the FAA in 1998 with over 25 years of service. I was fortunate enough to obtain a position as a flight and ground instructor with Ahart Aviation at the Livermore, California airport.

The 1992 *National Aviation Weather Program Plan* identified a number of unmet user needs. Among these was the necessity to improve aviation weather education for pilots and weather providers. This need was reiterated in the April 1997 *National Aviation Weather Program Strategic Plan*.

Aviation weather-related accidents continue to take their toll; analysis often reveals an inadequate, misinterpreted, or misunderstood weather briefing. With FSS automation, automated weather observations, and increased availability of direct and commercial weather information systems, pilots are required to read, interpret, and apply weather information on their own.

With all the books and articles written about aviation weather related accidents and incidents it might seem everything has been said; yet, I've discovered that very little practical information is available for pilots on this subject. Pilots, not only within the general aviation community, but all phases of aviation, have at their fingertips a wealth of information. However, few reports have an understanding of the information available to the pilot, either from FAA or nongovernmental sources; nor, do they present practical ways of translating, interpreting, updating, and applying information. My goal is to fill this need.

I am greatly indebted to many people for their generous help, guidance, and advice, too numerous to be listed in full. Among them are the meteorologists of the National Weather Service (NWS) at the FAA Academy in Oklahoma City, and local, regional, and national NWS offices, plus FSS controllers that I have been privileged to know, especially at the Oakland, California FSS. And, the pilots who have allowed me to assist them and in turn provided me with the best education possible. This book is dedicated to these people.

1

Introduction

Weather affects a pilot's flying activity more than any other physical factor. Most pilots agree that weather is the most difficult and least understood subject in the training curriculum. Surveys indicate that many pilots are uneasy with, or even intimidated by, weather. In spite of these facts, or maybe because of them, weather training for pilots typically consists of bare bones, only enough to pass the written test, and weather-related fatal accident statistics remain relatively unchanged.

Throughout the decade of the 1990s report after report, both government and industry, has recommended improved aviation weather education for pilots, dispatchers, controllers, and forecasters—weather users and providers. On September 27, 1997 Federal Aviation Administration (FAA) Administrator Jane F. Garvey issued the FAA's Aviation Weather Policy statement. It reads as follows in part:

> The FAA is committed to improving the quality of aviation weather information and the

application of that information by pilots, controllers, and dispatchers. The FAA acknowledges that training is a critical component of this objective, enabling the aviation community to make the best use of weather information to make sound operational decisions and to ensure safety and efficiency.

This policy was echoed by National Weather Service (NWS) Director John J. Kelly, Jr., in a letter written to the National Weather Association (NWA) in October 1998. Director Kelly supports the NWS's commitment to leadership and progress in aviation weather forecasting. He states:

Our aviation customers have expressed serious concerns about NWS aviation services. . . . New aviation products and services and improvements to existing products and services will be designed, validated and implemented. . . . Finally, I reaffirm the NWS dedication to the aviation program and I look forward to working with you to improve our aviation services.

The FAA, NWS, and the National Aeronautics and Space Administration (NASA), along with industry groups, such as the Cooperative Program for Operational Meteorology, Education and Training (COMET), have spent millions of dollars on aviation safety programs. However, very little has filtered down to the operational user or weather provider. To date, with the noted exception of NASA and the Aircraft Owners and Pilots Association's (AOPA's) Air Safety Foundation, little has been accomplished.

Accident reports and commentaries frequently refer to a pilot's poor judgment, namely, the failure to reach

Introduction 5

a sound decision. Pilot judgment is based on training and experience. Training is knowledge imparted during certification, flight reviews, seminars, and literature; experience can be best defined as when the test comes before the lesson. Unfortunately, failure can be fatal. Pilot applicants have only their instructor to prepare them to make competent flight decisions.

It's difficult to teach judgment—the ability to evaluate facts and come to a rational, safe decision. However, we believe that judgment, like common sense, is largely built on training and experience. Our goal is to share our training and experience. In this way we can, we hope, avoid the pitfalls of learning by experience. In other words: Learn from the mistakes of others; you won't live long enough to make them all yourself. The bottom line is that judgment and application of judgment, plus a knowledge of the aviation weather system and its relation to air traffic control, are essential to a safe, efficient flight.

When it comes to weather forecasts, Sir William Napier Shaw's *Manual of Meteorology*, published in the late 1920s, nicely sums it up: "Every theory of the course of events in nature is necessarily based on some process of simplification of the phenomena and is to some extent therefore a fairy tale." Are aviation forecasts accurate? The FAA admits that needs

> . . . cannot be met by an immediate application of existing technology. The need for accurate short-term forecasts exists in every phase of flight operations and is critical to an efficient, smoothly operating air traffic control system.

To this end the FAA and National Weather Service established Enroute Flight Advisory Service, including the implementation of high-altitude Flight Watch and Center Weather Service Units.

Each forecast is written for a specific purpose in accordance with specific criteria. Area Forecasts cover entire states, transcribed weather broadcasts (TWEB) Route Forecasts cover routes 50 miles wide, and Terminal Aerodrome Forecasts (TAFs) relate conditions basically within 5 miles of an airport. Differences are to be expected due to scale, interpretation of the weather situation, issuance times, and starting conditions. For example, localized areas of fog predicted in TWEB Route Forecasts or TAFs might not appear in the Area Forecast. Or, the Area Forecast might contain a prediction for thunderstorms that might not appear in individual TAFs when the forecaster does not expect the phenomena to occur at that airport. Forecasters might legitimately differ in interpretation. The Area Forecast might predict frontal passage at one time and the TAF might predict it at another time. Forecasts are issued at different times. Therefore, information available to the forecaster, on which to base the forecast, will differ.

A thorough understanding of format, limitations, and amendment criteria is required to adequately apply a forecast, especially when using a self-briefing media. The FAA and NWS have said,

> There probably is no better investment in personal safety, for the pilot as well as the safety of others, than the effort spent to increase knowledge of basic weather principles and to learn to interpret and use the products of the weather service.

Then there's the legal requirement. Each pilot in command is required by regulations to become familiar with all available information concerning a flight. This includes "for a flight under IFR or a flight not in the vicinity of an airport, weather reports and forecasts. . . ."

Pilots complain equally about pessimistic forecasts and unforecast weather. Pilots, often, have an over-expectation of forecasts. Each situation is different, with many variables and local factors. Forecasts are going to be missed. Errors fall into two categories: timing and the "Big Bust." The 1965 edition of *Aviation Weather* said it best: "The weather-wise pilot looks upon a forecast as professional advice rather than as the absolute truth."

Over the last decade there has been a revolution in the weather observing systems and in weather report and forecast format with the introduction of automated weather observations and the international METAR and TAF codes. In Chapter 2 we will review real and perceived limitations to automated weather reports.

Next we discuss low ceilings and visibilities and their effects on aviation operations. Chapter 3 includes strategies to preclude one of the most common mistakes—continued VFR into adverse weather. The proper and improper use of pop-up IFR operations are included.

Chapter 4 provides a discussion on nonconvective turbulence. Because of its operational significance we included an artificial form of turbulence—wake turbulence. The chapter includes strategies to minimize or avoid the effects of turbulence.

Chapter 5 continues with a discussion about the adverse affects of aircraft icing—induction, airframe, and carburetor, including strategies to minimize or avoid the icing hazard.

Thunderstorms contain just about every aviation weather hazard. Chapter 6 includes convective low-level wind shear, with all its perilous implications. Because of the significance of microbursts, a discussion of that topic is included. The chapter contains a comprehensive discussion of thunderstorm avoidance procedures.

Chapter 7 contains strategies for dealing with weather systems. Both frontal and nonfrontal weather systems are presented. Pilots need to understand that nonfrontal weather systems—surface and upper-level lows and troughs—present different hazards and require different strategies than frontal systems do.

In Chapter 8 we will explore misunderstandings and assumptions associated with density altitude and aircraft performance. These discussions include atmosphere affects on aircraft performance during all phases of flight.

Finally, Chapter 9 summarized the procedures pilots can use to avoid becoming involved in hazardous weather situations and weather-related accidents. We'll see how a string of relatively benign events has the potential for disaster and how to break the chain.

With respect to accident statistics and scenarios, some may wonder why we emphasize the negative. Our goal, along with AOPA's Air Safety Foundation and the FAA, is accident prevention. People learn either through training or experience. We don't want the learning experience to be where the test comes before the lesson. We hope, through these incidents and scenarios, to help prevent pilots from becoming statistics.

While we're on the subject of accident scenarios, let's concede that hindsight is always 20-20. It's easy to sit back in our favorite armchair and analyze and criticize someone else's performance and decisions. When we review an incident in this book, or any other publication or forum for that matter, let's not judge or attempt to blame. Our goal is prevention through education. Therefore, if the reader perceives any judgment or blame in any of these incidents or scenarios, it is strictly unintentional; that is not our purpose.

2

Automated Weather Observations

Most pilots are dealing with automated observations, to one degree or another, on a continuous basis. Like manual observations, automated observations have the potential to provide an extra degree of safety or lure the unsuspecting pilot into danger.

The new technology has its faultfinders. Flight Service Station (FSS) weather briefers, NWS forecasters, weather observers, and pilots are among the critics. There is no question that automated observations, like anything new, have had "teething" problems. The two most controversial elements are visibility and sky condition. Critics cite these element as the least accurate; they are certainly the least understood. Additionally, accommodation must be made for the lack of certain weather sensors. But, we'll find in the final analysis that most criticism is due to misunderstanding, a reluctance to change, or personal prejudice.

Visibility

Automated systems determine visibility from a scatter meter device. The visibility sensor indirectly derives a value of visibility corresponding to what the human eye would see. The visibility sensor projects light in a cone-shaped beam, sampling only a small segment of the atmosphere—an area about the size of a basketball. The receiver measures only the light scattered forward. The sensor measures the return every 30 seconds. A computer algorithm—mathematical formula—evaluates sensor readings for the past 10 minutes to provide a representative value. Reported visibility is the average 1-minute value for the past 10 minutes.

One misunderstanding is that automated machines extrapolate prevailing visibility based on sensor data. (Prevailing visibility: Visibility reported in manual observations—the greatest distance that can be seen throughout at least half the horizon circle, which need not be continuous.) This is not true. Automated visibility is not prevailing visibility and may be considerably different. The existence of fog banks and visibility in different sectors will not, normally, be reported. Automated visibility may not be representative of surrounding conditions. Regulations specifically take these variables into consideration for IFR, where a suitable alternate may be required. VFR pilots must use the same caution and, although not specifically required by regulations, plan for an acceptable VFR alternate during reduced visibility or when conditions are forecast to change.

During rapidly decreasing conditions it takes between 3 and 9 minutes for the algorithm to generate a SPECI—a special surface aviation weather report. During rapidly increasing conditions the algorithm takes between 6 and 10 minutes to catch up with actual conditions. This feature adds a margin of safety and buffers rapid fluctuations in visibility.

Siting the visibility sensor is critical. If the sensor is located in areas favorable for the development of fog or blowing dust or near water, it may report conditions not representative of the entire airport. Airports covering a large area or near lakes or rivers may require multiple sensors to provide a representative observation. To this end, some airports will be equipped with more than one visibility sensor. Site-specific visibility, which is lower than the visibility shown in the body of the report, will appear in remarks (e.g., VIS 2 RY11; visibility two, at runway one one).

CASE STUDY Overheard from a Stockton, California, approach controller: "Visibility minus one-quarter." Can visibility be less than zero? I don't think so. The controller must have meant "less than one-quarter."

Pilots and controllers have criticized automated observations because of the perceived frequency that visibilities are too high. Some pilots routinely cut the visibility in half when light precipitation is falling in the area. In the presence of rain, snow, drizzle, and fog, considerable errors have been reported. Automated systems tend to be overly sensitive to ground fog. Under these conditions the human eye is affected by bright back-scattered light, which sharply reduces visibility. This is comparable to the headlights of a car shining into fog or snow. The brightly reflected light may blind the observer and limit perceived visibility. Yet the lights of an approaching vehicle seem to penetrate the fog; an observer can see the approaching light source farther into the fog. This is caused by light reflected back toward the observer—forward scattered. Research shows that visibility differences under these conditions between forward- and back-scattered light are on the order of 2 to 1. However, there are no guarantees. For example, in low-visibility conditions at Fresno and Bakersfield, in California's San Joaquin Valley, tower visibility has been reported greater than that reported by automated surface observing system (ASOS). Some of this difference can be attributed to the difference in observation sites between the ASOS sensor and the tower.

If conditions are bright enough to use sunglasses, expect the automated systems to report visibility about twice what the human eye perceives. At night, human observers report visibility using forward-scattered light, the same principle used by automated sensors. Therefore, manual and automated visibilities tend to be more consistent.

Manual and automated visibility may be considerably different, especially during the day. The existence of fog banks and visibility in different sectors may not be reported. Reported visibility, manual or automated, may not be representative of surrounding conditions. Reports from surrounding locations, pilot reports (PIREPs); and satellite images can help fill in the gaps. The bottom line is that with manual or automated reports close to visual flight rules (VFR) or instrument flight rules (IFR) minimums, pilots must exercise additional caution, carry extra reserves of fuel, and have a solid alternate.

Both manual and automated observations face physical limitations, such as site location, rapidly changing conditions, and contrast.

CASE STUDY As reported: the ASOS in Kalispell, Montana, was reporting visibility $1/4$ mile. I was given a clearance to descent to 8000 ft and hold at the Smith Lake NDB [nondirectional radio beacon]. I cancelled IFR and informed air traffic control (ATC) that I was going to the airport to see if the ASOS was correct. I proceeded VFR. I have seen this system report inaccurately in the past. I flew to the runways and ascertained that runway 20 was clear for approximately one-half to three-quarters of its length. There was fog over the approach end of runway 2 (runway 2 is the instrument-landing system, or ILS runway and the location of the ASOS). I remained VFR and landed on runway 20.

The reporter states that the ASOS was incorrect. At least in this case, that's not true. The ASOS was perfectly correct. Had a human observer been in the same location as the ASOS, a manual observation would have reported the same conditions.

The reporter went on to say, "My contention is that the weather system is inadequate and unsafe in its present condition." ASOS bashing? The ASOS was entirely accurate within the scope and limitations of the equipment. Certainly, the weather report required this flight to have a suitable alternate airport and fuel reserves. Had the whole airport been fog covered, the pilot would have simply proceeded to the alternate.

Ironically, the reporter went on to say: "At no time was safety compromised. My landing was a judgement call. I was never out of VFR conditions."

Let's not forget the VFR pilot. The ASOS report would alert the pilot to have adequate alternates and fuel reserves, which, more than likely, would be greater than that required by regulations. Did the reporter provide a PIREP of existing conditions? It certainly would have been very helpful to ATC, forecasters, and other pilots.

CASE STUDY As reported: ASOS was not representative of actual weather conditions. Tower visibility was 10 miles, with some patchy ground fog. The ASOS was reporting 2 miles. Tower personnel contacted the contract observer, who would not augment the erroneous report.

The reporter commented: "While ASOS may do a great job augmenting the . . . area where the equipment is physically located, . . . it does not report conditions at, or near the airport accurately." Here, again, we may very likely have a problem of different conditions existing over different parts of the airport.

The contract observer has overall responsibility for the observation. If the observation is provided by a contractor for the FAA, the pilot can make an official complaint to the FAA's Air Traffic Manager. This should be done within 15 days of the occurrence. The information in this ASRS report is certainly not sufficient. The reporter must provide enough information, as detailed as possible, of the occurrence to the FAA. The FAA can then investigate the incident to determine if, in fact, the contract observer's performance was unsatisfactory and take appropriate corrective action.

Atmospheric Phenomena

Atmospheric phenomena is weather occurring at the station and any obstructions to vision. Obstructions to vision are normally only reported when the visibility is less than 7 miles.

CASE STUDY As reported: An intensity 5 thunderstorm was 10 miles west of the airport moving east, toward the airport. Clouds associated with the cell created an overcast layer above the airport. ASOS reported the weather as visibility 10 miles, clear. This was not very representative of the weather at the time.

Many ASOS sites are not yet equipped to report thunderstorms. This is typically indicated in the remarks of the observation as ". . . RMK TSNO." If the clouds were above 12,000 ft, the ASOS would report ". . . CLR . . . ." Pilots, dispatchers, and FSS briefers need to understand the limitations of all weather reports and forecasts; automated observations are no exception. As required by regulations, all available information must be analyzed and applied to every flight.

A major criticism of automated observation is its inability to detect freezing precipitation or thunderstorms. Steps are currently being taken to address these limitations. Automated sites are being equipped with freezing precipitation sensors. This will allow automated stations to report the occurrence of freezing rain and freezing drizzle. The Automated Lightning Detection and Reporting System (ALDARS), which acquires lightning information from the National Lightning Detection Network, will allow automated weather observing system (AWOS)/ASOS to report the occurrence of a thunderstorm. ALDARS is operational at numerous AWOS sites and is expected to become operational with all of the FAA's commissioned ASOSs.

Automated reports can and should be supplemented with radar products. The system is now virtually complete and covers most of the country. Radar can determine the existence of rain, thunderstorms, tornadoes, snow, and hail. Radar, along with satellite imagery, is better at determining the extent of phenomena than either a manual or automated observation.

Sky Condition

Automated stations determine sky cover and height from a laser cloud-height indicator (CHI). Similar to a rotating beam ceilometer, cloud elements reflect the laser. Like visibility, a computer algorithm processes the last 30 minutes of CHI data. The computer then generates values of sky cover and cloud height for the observation. To be more responsive to the most recent conditions, the algorithm "double-weighs" the last 10 minutes of data. Up to three layers are reported.

At the transition between scattered and broken sky cover, human observers often report too much cloud cover. This is known as the "packing effect," a condition where an observer does not detect the opening in the cloud deck toward the horizon. Pilots also tend to overestimate the amount of cloud cover. ASOS is not biased by these limitations.

In rapidly changing conditions, the automated system algorithm tends to lag slightly behind actual conditions. If a sudden overcast layer develops, ASOS will take 2 minutes to report a scattered layer; within 10 minutes the system will report broken conditions.

FACT? Sky cover is always an estimate from a manual station. As an FSS briefer for some 25 years, I have on occasion briefed doom and gloom only to find a bright, beautiful day. It would seem that certain tower controllers use the following criteria: They consider the roof of the tower cab as opaque; therefore, one cloud is scattered, two is broken, and three is overcast with breaks.

On rare occasions, ASOS may report a dense moisture layer as clouds before the layer becomes totally visible to the eye. This may occur with an approaching cold front when the sensitive laser detects the large-scale lifting of prefrontal moisture. There have been cases when ASOS reported a layer 20 minutes before a human observer did.

ASOS will only report conditions that pass directly over the sensor. During light wind conditions, observers have reported up to three-eighths sky cover when ASOS reported CLR. Ironically, manual observations suffer from the same limitations. Sky cover might not be representative of surrounding conditions, especially at night or during low visibility, when the observer cannot see or evaluate the whole sky.

Manual observation often suffers from many of the same limitations as automated observation. For example, sky cover and cloud height might only be an estimated, a more or less educated guess by the observer, based on the observer's training and experience. Like visibility, cloud cover and heights—manual or automated—should always be viewed with caution, especially at night or close to minimums.

FACT The FAA has conducted "blind" comparisons between manual and automated observations. At many of these locations the ASOS was installed but not commissioned. A number of the observers complained when they were told they would not be able to use the laser CHI for cloud heights, which they had been previously using to supplement their observations.

At certain automated locations additional sensors are used to obtain more representative reports. In such cases, remarks will identify site-specific sky conditions that differ and are lower than those reported in the body. For example, . . . RMK CIG 020 RY11 . . . , ceiling two thousand at runway one one.

CASE STUDY As reported: The pilot tuned in the AWOS and received a report of 100 scattered. Over the final approach fix the pilot experienced a solid undercast with tops at 1300 ft. There were no suitable VFR airports in the area. The pilot, unable to land, was forced to return to the departure airport.

The AWOS was reporting clear. On approach an air carrier stated: "Unable to make a visual approach because of clouds. The airport is the only clear spot in the area."

Both of these case studies illustrate the need to obtain all available information. If all other airports in a general location are reporting overcast, the pilot should view a report of 100 scattered with some skepticism. On the other hand, the airport being the only clear spot in the area is not uncommon. If we have the capability of flying IFR, stay on the IFR flight plan until landing VFR is assured. We don't know if the satellite picture, PIREPs, or forecasts were obtained by these reporters. They might have given the pilots additional insight.

The reporter in the first incident commented: "This AWOS error could have caused a serious fuel problem for a long-range flight arriving with minimum required fuel remaining." For IFR operations, regulations take these types of problems into consideration by requiring fuel to alternate airports, plus an additional fuel reserve. Minimum fuel reserves for VFR flights are just that—minimum. Minimum does not necessarily equate to safe. As soon as an undercast is encountered, a VFR pilot must check to determine its extent. It makes no sense to fly for 45 minutes over an undercast, betting your life that an airport will be clear with only a 30-minute fuel reserve.

It might be a case of a faulty sensor. When verified, the FSS will issue a Notice to Airmen (NOTAM) indicating that an individual sensor or sensors are unreliable. The FSS or controlling agency will also notify maintenance personnel. Like the case of the contract observer failing to correct a nonrepresentative report, we as pilots have an obligation to report faulty equipment.

Manual versus Automated Observations

In 1997, at the direction of Congress, the FAA conducted an ASOS Operational Assessment. The report concluded that, overall, ASOS performs as designed. ASOS required edition/augmenting 16 percent of the time for "nonrepresentative data." Cloud height, cloud coverage, and visibility accounted for the majority of edits. "At no point did the team consider any of the concerns significant enough to curtail the development of future ASOSs, nor did any of the concerns impact safety, efficiency of operation, or airport capacity. . . ." ASOS availability was excellent. Sensor reliability exceeded 99 percent. "The ASOS reporting of changing cloud heights and sky coverage was the area in which most of the inconsistencies (whether perceived or real) were found. . . . ASOS lagged behind the observer in reporting lowering ceilings and led the observer in reporting rising ceilings."

CASE STUDY During a survey conducted at the Oshkosh Fly-in in 1997, 568 pilots responded to this question: How often do the conditions disseminated by AWOS/ASOS match those which you experience in flight? Forty-five percent responded "always" to "often;" another 19 percent responded "occasionally." Only 5 percent responded "never" or "infrequently." The remaining 35 percent indicated "don't use" or did not respond. This would seem to refute the notion that automated observations are less accurate than human reports. In fact the survey could be interpreted to mean that automated reports are as accurate or more accurate than human observations.

I must point out that human observers average visibility, sky condition, and cloud height, typically over a period of 10 to 30 minutes, just like ASOS. The addition of reporting locations and availability of weather is an undeniable advantage of automated systems. For example, a pilot might cancel a flight rather than "taking a look" with IFR or marginal VFR reported. A pilot in flight could divert before encountering severe conditions, rather than possibly arriving without enough fuel for a suitable alternate.

The need to evaluate all available sources cannot be overemphasized—adjacent weather reports, PIREPs, radar reports, satellite imagery, and forecasts. A sound knowledge of aviation weather can help with evaluation and identify erroneous observations. For example, on December 1, 1995, Fresno, California, reported a tornado. The synopsis forecast high pressure, radar showed no precipitation, and the satellite indicated no cloud cover. So what happened? The observer, augmenting the report, was practicing entering supplemental data—which was inadvertently transmitted. Other manual and sensor errors can be detected in this way. Whether it is a manual or automated observation, pilots must remember that reported conditions may not be representative of the surrounding area.

As an old aviation axiom laments: Aviation weather reports may not be accurate, but they're official. The question is not whether manual or automated reports are better or worse. They are different, with both having advantages and limitations that must be understood.

CASE STUDY As reported: Upon taxiing out for a VFR departure, the weather was reported by a manual observation as visibility 3 miles, ceiling 1400 ft broken. After departure the pilot encountered lower clouds.

The reporter commented: "There is a 15-minute delay between the observation of the weather and the time ATC gets the weather. It's old news. Looking forward to ASOS so that you can get current and accurate weather that's up to the minute—not 15 to 30 minutes old." This comment is certainly true with most automated observations and is one of its greatest advantages.

Perceived Accuracy

CASE STUDY As reported: The pilot of an air carrier asked the controller for the current weather. The controller replied 1 mile, moderate freezing rain, 800 overcast. The pilot argued, stating they were not getting any precipitation. The pilot asked for the weather observer to take an observation. The observer verified the report.

This report was filed by a frustrated air traffic controller. It seems in the reporter's experience that ". . . it gets to be a problem when the carriers don't believe the ASOS machine, let alone the weather observer. We as weather observers can't change the weather just to accommodate the air carriers."

This situation is nothing new. It's been going on since weather observations and regulations began. Pilots must understand that weather can be different over different parts of the airport. Controllers must understand that weather has a significant effect on aircraft operations. In this case the crew was concerned about having the aircraft deiced again. The reporter went on to say, "There should be a familiarization program set up for them to visit ATC facilities and see what we do." This is a two-way street. Pilots are encouraged to visit ATC facilities. However, the FAA has done little, if anything, to familiarize controllers with the operational needs of pilots.

A callback conversation with the reporter revealed the following: "The reporter felt that the ASOS usually provides very accurate data and finds it unnecessary most of the time to augment observations."

CASE STUDY As reported: Unforecast ground fog moved in over the destination airport, reducing visibility to $1/4$ mile. I asked the captain what he wanted to do. He immediately questioned the reliability of the ASOS. He wanted to fly over the airport and take a look. I informed him that we had only two options: hold until the ASOS visibility improved or go to our alternate. I insisted that an approach to "have a look" was not only unsafe, but illegal. He opted to hold. ATC informed us that the last weather report prior to the tower closing—20 minutes earlier—was visibility 2 miles. The captain decided it was now legal to make the approach. I was not comfortable with this but resigned myself to his decision. We could see the airport straight down through the fog. At 100 ft we could not see the runway lights, and executed the missed approach.

In this incident ASOS did its job. The reporter commented that ". . . in hindsight, I feel that accepting the approach was clearly the wrong decision. The issues I considered are as follows: The ASOS weather was more current and is our approved source of weather."

CASE STUDY As reported: The forecast was for a ceiling of 2500 ft. The pilot departed and entered clouds at 600 ft above ground level (AGL). The ASOS was operating in a test mode that indicated IFR conditions. A plane had just landed VFR and the ceiling looked VFR to me, so I decided the ASOS must be inaccurate.

The reporter, based on the above information, decided to depart VFR and pick up an IFR clearance enroute. It appears the pilot's decision was based on the need "to get going" rather than an objective evaluation of conditions. Unfortunately, with the information available it's difficult to do a more detail evaluation. However, the reporter did comment: "I should have gotten a clearance on the ground." This would have certainly been the safest option based on the prevailing conditions.

3

Low Ceilings and Visibility

An all-too-frequent causal factor in aviation accidents is continued VFR flight into adverse weather. However, armed with knowledge of these phenomena, a VFR pilot can avoid the hazard and nullify the risk. The IFR pilot is also concerned with ceiling and visibility, especially in the takeoff, approach, and landing phases. No segment of aviation is immune. Take, for example, the following case study.

CASE STUDY A U.S. Air Force T-43 (Boeing 737) crashed in low-ceiling and low-visibility conditions in Bosnia, killing all aboard, including Secretary of Commerce Brown.

Ceilings and Visibility

Of the pilots involved in low-ceiling accidents, more than half had instrument ratings. Also, more than half of the accidents were fatal.

CASE STUDY The pilot departed a Southern California airport during the evening for a flight to Monterey. Arriving in the Monterey area about midnight, the pilot found the airport overcast. The pilot landed on a highway south of the Reid-Hillview Airport in the San Francisco Bay area.

There was no record of the pilot checking enroute for a weather update, which would have revealed the onset of coastal stratus. There is no rational reason for this accident to have occurred. The pilot landed safely and was applauded by some. Unfortunately for aviation, opponents of Reid-Hillview cited it as another reason to close the airport.

In the preceding example even the required fuel reserve might not have been adequate. A 45-minute fuel reserve doesn't make any sense with the nearest suitable alternate 50 minutes away. What would have happened if the coastal stratus extended beyond the airplane's range? Most likely the accident would have been fatal. How could this incident have been prevented? Simple: update weather enroute with Flight Watch or Flight Service.

When conditions are favorable for coastal stratus, VFR pilots should plan arrivals and departures during the afternoon hours. If this is not possible, moving the aircraft to an airport a few miles inland will often allow a morning departure.

CASE STUDY As reported: I had departed Gallup, New Mexico, for Scottsdale, Arizona. I proceeded southwest toward better weather conditions than the direct route I had filed for my VFR flight plan. As I turned to a southerly heading, it became clear to me that continued VFR toward Scottsdale was improbable. I called center and picked up an IFR clearance into Sedona, Arizona. As I climbed to my assigned altitude, the aircraft started to pick up ice. At this point I elected to cancel IFR and descend back to a lower altitude. I could see down through broken clouds to good visual ground contact. I turned northeast toward Winslow, Arizona, to find that continued VFR flight was, again, improbable. I was left with lowering clouds, darkening skies, and fuel becoming a factor. I picked up a southerly heading and after several minutes contacted center for radar vector into Payson, Arizona. After switching to a different controller, the controller was reluctant to help because of my low altitude and his limited radar coverage in that area. I finally insisted upon suggested radar headings to find the airport. Using the intermittent radar returns, the controller was able to give me headings to the airport.

The ASRS report indicated the pilot had considerable experience, including flying jet aircraft. The pilot commented, "I have never had a more challenging flight than that evening up on the Mogollon Rim northeast of Phoenix."

Could the pilot, with vast experience, have been complacent? Certainly the pilot pushed on into, admittedly, improbable VFR weather. The pilot did not know the airplane's position, having to insist that ATC provide radar vectors to an airport. The pilot commented, "The springtime weather may be OK in Phoenix and Gallup, but the weather up on the Rim may not be. I do not think I broke any rule, but I'll never do this again."

Did the pilot obtain a weather briefing for the flight as required by regulations? The pilot, apparently, was not aware that weather conditions can change drastically, especially in mountainous regions. Did the pilot maintain situation awareness of the airplane's position? What would have been the result if the pilot were unable to obtain radar assistance? Finally, in the event of an accident, what would have been the National Transportation Safety Board's (NTSB's) probable cause?

CASE STUDY As reported: I checked the weather before I left Palo Alto, California, and the Central Valley was a mess with haze, but Southern California was forecast to be fine for VFR flight into Fullerton. I followed I-5 from the Grapevine to Castaic. At this time it was obvious that the entire Los Angeles basin was covered by a layer at 3100 ft. I told the controller that I would like to follow I-5 to Fullerton. The controller said that might be possible at 2000 ft. I said fine and responded that I saw a hole about 10 miles ahead. When I dropped down, it was low visibility, but not technically IFR (3- to 5-mile haze). Having no desire to fly in these conditions, I climbed back to 7500 ft. I headed northeast to Lancaster and landed.

Like so many of the case studies, there is no evidence of the pilot checking weather with Flight Service or Flight Watch enroute. The reporter commented, "When I saw the cloud layer, my first thought was to return to Bakersfield." This whole incident would have been avoided if the pilot had gone with that thought. Countless accidents and incidents could be avoided if the pilot would turn around when that thought first occurs. Unfortunately, it's human nature to push on to get home. This behavior, however, can be changed but only by a conscientious effort on the part of the pilot. Like any addiction, the first step to a cure is admitting that there is a problem. These aren't called "sucker holes" for nothing. This pilot was fortunate to be able to file an ASRS report rather than an accident report.

CASE STUDY The pilot reported that a "grossly inaccurate weather forecast caused a planned VFR arrival at an unfamiliar airport to be made when surface-based Class E airspace was IFR due to low ceilings." The pilot reported being confused about the proper controlling agency for the airspace and that UNICOM [uniform communication] was unresponsive. The pilot admitted "insufficient planning for an alternate landing outside of controlled airspace."

It's difficult to assess what the pilot means by a "grossly inaccurate weather forecast." Certainly the pilot failed to update weather enroute. The pilot was apparently untrained in determining the controlling agency for surface-based Class E airspace and the scope and purpose of UNICOM.

By regulation, pilots are required to be trained in the "procurement and analysis of aeronautical weather reports and forecasts, including recognition of critical weather situation and estimating visibility while in flight." The pilot was unaware that the "controlling agency" can be found in the *Airport/Facility Directory* or by calling the Flight Service Station or that UNICOM is a nongovernment communication facility that may provide airport information at certain airports. The pilot admitted not planning for an alternate. The pilot, presumably, came from an area of VFR weather, but blindly (pardon the pun) proceeded on.

CASE STUDY A Navion pilot requested and received a special VFR clearance from the Salinas, California, tower out of Class D airspace. The pilot then reported clear of Class D airspace. The airplane crashed 8 nautical miles from the airport in rising terrain. At the time of the accident the weather was reported as KSNS 151647Z 29010KT 5SM BR OVC004 15/13 A2996.

This pilot was legal for weather minimums but not for minimum safe altitudes. Unfortunately, this pilot will not be able to represent himself at the NTSB hearing; he was fatally injured.

Continued VFR

Continued VFR into adverse weather is an all-too-common probable cause. It is a sad commentary that all of these accident are preventable.

CASE STUDY As reported: About 1 hour before the flight I received a weather briefing for my route. The route was reported as VFR, except for some fog over Jackson.

Jackson was about 30 miles north of my course. I called back to file my flight plan. After takeoff I had opened my flight plan. During climb I noted the city lights of my first check point. It was about 1 hour before sunrise and the visibility was good. About 45 minutes into the flight, while looking down for my next checkpoint, I noticed that the ground lights began to flicker. When I looked up, the windscreen was gray. When I looked down, all the lights had vanished. I was in a cloud.

I had not expected to encounter this cloud, but there it was. I noticed some small patches of ice had formed on the windscreen. The visibility, or lack thereof, at this point seemed the least of my problems. I assessed that, according to training, I should climb to a higher altitude. That, as I recall, was preferred to descending, although each would probably work. If, however, I climbed and exited the cloud, I would still not have ground reference. I was thankful for a little over 3 hours of instrument training.

At this point I began a 180° turn, yet I remembered that I had flown through something that deposited ice and did not want to reencounter that event. What to do? I called Flight Service and reported that I had no visibility with ice on the windscreen. I also remember saying that I was only VFR certified. I asked for a report on the extent of this cloud and which direction would be appropriate for me to exit and return to VFR conditions. They instructed me to descend to 2500 ft. I followed instructions and found that once again I could see the ground. I was next instructed to contact approach control. The flight continued uneventfully to destination.

This pilot was very fortunate to be able to file an ASRS report rather than being involved in an accident—which, based on statistics, would have been fatal. The pilot's decision to climb was not appropriate. It appears this was based on the pilot's interpretation that when icing is encountered to climb. This may be fitting for an

IFR pilot, but in this case the proper procedure would be to initiate immediately a 180° turn.

CASE STUDY As reported: I was flying at 600 to 700 ft AGL beneath ceilings of 1000 to 1500 ft when I encountered clouds. I suppressed a nearly irresistible desire to fly lower and get underneath. This was clearly wrong, for all I knew this wasn't a cloud, but fog extending to the ground. I executed a standard rate turn for 1 minute, noted that my heading was 180° opposite the previous heading, and exercised patience during 2 or 3 unending minutes before returning to marginal VMC conditions. I landed safety, concluding the flight.

After inadvertently entering clouds, the reporter made two sound decisions. The first was not to descend. Because cloud bases were unknown, this action very likely could have led to disaster. The second sound decision was to execute a standard rate 180° turn to the reciprocal heading and then continue on that heading. The reporter than describes 2 or 3 unending minutes. In any anxious situation every minute seems like 10. The reporter's third sound decision was to have patience while exiting the cloud.

The reporter went on to list the following contributing factors:

1. A desire to get home
2. Darkening sky, making overall visibility worse
3. Fatigue, from having flown all day
4. Previous experience flying in marginal conditions
5. An inappropriate level of optimism given the situation

This case study pretty much speaks for itself.

CASE STUDY As reported: The flight instructor obtained a weather briefing indicating VFR conditions. After departure we entered a fog bank at 800 ft. We tried to stay below the clouds at 700 ft as we were in and out of fog. Abeam the end of the runway we were back in VFR weather. It was night and we couldn't see the fog. Another aircraft had taken off prior to our departure and did not report fog or low ceilings.

Recalling the discussion in Chapter 2, manual or automated weather observations, especially at night, may not be representative of surrounding conditions. This case study emphasizes the need for PIREPs. Had the previous pilot reported conditions, this incident very likely would not have occurred.

CASE STUDY As reported: I wanted to move my plane this afternoon. While driving home I could see above the tops of the hills. The weather called for snow this weekend, and I figured that I would not be able to get the airplane home before the snow hit the ground. With a 3000-ft ceiling, I blundered off into the skies with lower ceilings ahead. Five minutes into the flight I looked at the compass and said to myself that 180° from this is my way out. I was 5 miles from my destination, and thought that I could just follow the road into my home strip. This was the biggest mistake of my life!

For about 5 to 8 minutes I was totally disoriented and struggled to keep the plane right side up. I said to myself that this is it! After 4 to 5 minutes of total IMC [instrument meteorological condition(s)], I finally got somewhat of a grip. I climbed up to 2400 ft and prayed. I also headed in the direction which was my escape. I was in and out of the clouds when I saw my home base through a hole. I taxied to my tie-down and said a prayer to my God that I was still alive.

The reporter went on to say, "I do not write this as an adventure/thriller but to help others not to make my mistake! I could easily have been on the NTSB list of accidents. It is not funny or heroic by any means. I was stupid. Do not scud run or fly into IMC—period! If all else fails, remember what you were taught—it saved my life."

Low Ceilings and Visibility

The reporter makes some very valuable points. First, don't let yourself get into this situation. If you do, don't give up. Fly the airplane. The reporter also mentioned switching to 121.5 for assistance but then was fortunate enough to find a hole. Pilots should never hesitate to request assistance; that is a primary function of ATC. It cannot be overemphasized that you should obtain help before an incident becomes an accident. Remember the axiom: You want to be able to represent yourself at the NTSB hearing.

CASE STUDY As reported: The weather was poor. I continually called flight service to look for improvement. The weather to the south was gradually improving, but the destination remained socked in. I was anxious to go in order to get back to my business. Unfortunately, I decided to fly despite the fog, figuring I'd climb on top and descend close to home. At that time the destination had a 1900-ft ceiling. Upon climb into the fog, I became disoriented. I immediately called ATC to let them know what was happening. ATC vectored me through the clouds.

The reporter stated, "My decision to forgo weather in order to get back to my business was, to say the least, irresponsible. I respect weather, but my business sense told me otherwise. The fact that I lived through this experience brought home just how dangerous weather can be, if one chooses to fly in it. My trouble inflight was only a few seconds, but my memory of it will last my lifetime."

CASE STUDY As reported: I was flying on a pleasure flight VFR. I was between layers of solid clouds. The layers converged. I pressed on because my destination was only 12 miles away, reporting broken skies. The airplane iced and so did the pitot. I applied pitot heat and turned 180°. Pitot instrument function soon returned. I broke out into VMC in approximately 10 minutes and returned home.

These, and similar, incidents brings home the point that the minimal instrument training required for the private pilot certificate can, often, become a double-edged sword. The goal is to allow a pilot to extricate him- or herself from an inadvertent encounter with clouds. Unfortunately, many pilots use this training to knowingly penetrate clouds. (I know it did in my young and inexperienced days.) Every year pilots insist on pushing the envelope, all too often with fatal results. It cannot be overemphasized that the correct behavior is the retreat from below-VFR conditions before they are encountered.

Pop-Up IFR

We have an extremely flexible ATC system. It allows a pilot to simply call the sector controller and obtain an IFR clearance. Unfortunately, many pilots rely too heavily on this procedure. ATC cannot always accommodate such requests.

CASE STUDY As reported: We were on an IFR flight plan, but as we neared our destination at our cruising altitude of 6000 ft, it began to clear. The NDB at our destination was out of service, but the forecast was for ceilings of between 1500 and 2000 ft. Flight Watch confirmed that weather reports near our destination were VFR. A commuter flight in the vicinity indicated that there were a few clouds at 1500 ft and that the ceiling was about 3000 ft. Knowing that the minimum vectoring altitude (MVA) in the area was 4000 or 5000 ft, and knowing that I would be unable to do the NDB approach, I knew that our only chance at getting in would be to proceed VFR under the overcast. Just northeast of our destination was a large hole in the clouds, through which I could see the ground. I decided to cancel IFR and descend VFR.

I descended and got below the clouds at 2000 ft MSL [mean sea level] (600–700 ft AGL [above ground level]). I was able to maintain VFR in Class G airspace. The weather had begun to deteriorate and conditions were worse than both the forecast and the PIREP. As I headed toward the destination, I realized that I would be unable to maintain VFR for much longer. I radioed ATC and requested another IFR clearance and a climb. I was told to stand by. At that time, I entered IMC. Because I was so close to the ground and conditions had changed so rapidly, as I waited for my clearance, I decided to start a climb away from obstructions. At this point, I figured that climbing straight ahead into IMC was safer than attempting a low-altitude 180° turn back to conditions I was unsure were still VMC. ATC called back with a clearance and we landed at another airport.

The pilot commented, "While it would have been best not to have ever gotten myself into the situation in the first place, once presented with the situation, it was safer to climb rather than attempt a low-altitude 180° IMC turn over unfamiliar terrain." Certainly a factor that compounded this incident was the pilot's unfamiliarity with the area. Without some positive indication (AWOS, visual contact with destination, satellite image, or PIREP for the destination) that the destination is VFR, it usually isn't a good idea to attempt to "scud run" to the destination. The pilot further remarked, "In the future, I realize that it will be better to land at an alternate airport, served by an operational approach, than to attempt to fly under clouds, in marginal conditions, to my intended destination."

CASE STUDY As reported: On a cross-country flight the weather began to deteriorate, so we called ATC and requested an IFR clearance. After repeating the request no less than four times, and holding for 20 to 25 minutes, ATC had not given a clearance or a squawk code. At this time low cloud cover appeared to be clearing. We continued. However, the ceiling got lower, and in about 20 miles we went IFR without a clearance. At this time, we called ATC and got an IFR clearance with no delay. The flight continued uneventfully.

The reporter commented, "I believe that not getting services from ATC and the wish to press on caused this to happen." ATC's mandate is the separation of known IFR aircraft and the efficient flow of air traffic.

Except in an emergency, pilots cannot expect ATC to drop everything to service their needs. What options did this pilot have—that is, other the blundering into the clouds? If conditions are marginal, plan an IFR flight from the point of departure. Update weather with Flight Watch, and if there is a problem ahead, get into the system before adverse weather is encountered. This may mean filing an IFR flight plan with flight service enroute. Another option would be to file an IFR flight plan to be picked up enroute, well before entering marginal conditions. The point is to get into the system.

CASE STUDY As reported: I was dispatched, but in order to expedite the flight, I did not file an IFR flight plan. My first destination was 15 miles from my departure point. I contacted ATC and was handed off to several controllers before finally receiving a squawk code and a clearance to enter Class B airspace. After entering IMC, I was assigned several vectors and altitude in order to get the ILS to my destination. After failing to maintain the proper headings, ATC asked me if I was in trouble and if I needed any special assistance. It was not until then that I realized that my radio magnetic indicator (RMI) was inoperative. I tried to troubleshoot in order to rectify the problem, but I became task saturated on vectors while trying to reach for the proper approach plate. I became disoriented and was unable to maintain assigned headings and altitude in this busy airspace.

A callback conversation with the reporter revealed the following information: The failure of the RMI was caused by a tripped circuit breaker not powering the indicator. The reporter said maintenance reset the breaker and the system operated normally.

This incident reveals a series of relatively small occurrences that led to the problem. The pilot was in a hurry. Because there was no time to file a flight plan, it appears there was not time to properly preflight the aircraft. Had the trouble with the RMI been discovered on the ground, it would have been corrected or the flight canceled. The pilot was hurried by ATC, again a result of initially not being in the system. Numerous vectors and altitude and frequency changes are often the result of ATC doing their best to fit pop-ups into the system. This is especially true in congested Class C and Class B airspace. It appears poor cockpit management contributed to the incident. The pilot reported having to search for the proper approach plate.

CASE STUDY As reported: The weather had changed to IFR conditions up through Albuquerque. However, the route west through El Paso was reported VFR with isolated rain showers. Ceilings at Odessa were reported at 2100 ft AGL and improving along the route to broken to overcast 4500 ft at Salt Flat and clear at El Paso. Not wanting to fly an unfamiliar plane IFR, I chose the route through El Paso, using flight following in lieu of a flight plan. The departure out of Odessa was uneventful with adequate ceilings to maintain level cruise at 4500-ft MSL with a 1500-ft cloud base separation. Thirty miles west of Wink the ceiling lowered to 1500 ft AGL, so we descended, lost contact with flight following, and shortly thereafter encountered snow obscuring forward vision, leaving only vertical vision to the ground. At this low altitude the VOR [very high frequency omnidirectional radar] was unreliable. Along the route we passed several tall antenna, thus a decision to turn around at the low altitude did not seem prudent.

I called ATC for an IFR clearance, climbed to 10,000 ft, and picked up ice during the climb. We broke out on top at 10,000, but chose to climb to 10,600 to maintain clear of clouds on top. The rest of the flight was uneventful.

Unfortunately, the report does not indicate if the pilot obtained a standard weather briefing prior to departure. The pilot elected to fly low under the clouds, even though the airplane was equipped for IFR. This resulted in the pilot become trapped in low-visibility conditions. The pilot reject a 180° turn because of several tall antenna. The pilot reported being unable to see ahead, but decided that a decision to turn around at the low altitude did not seem prudent. Could this have been a excuse to press on? After losing radar services, the pilot reports obtaining an IFR clearance. It appears some flight was conducted in less-than-VFR conditions. The pilot was very fortunate to be able to obtain a clearance. This is uncongested airspace and receiving a pop-up clearance is more likely than in other airspace.

A more prudent, and certainly safer, procedure would be to retreat to an area of good weather and either land or file to obtain a clearance. Was the pilot aware of cloud tops? If the tops had been higher, this could have very easily ended up as an icing accident. The airplane was not certified for flight into known icing.

The reporter stated: "Solution? Better weather information along route of flight. Since weather is hard to anticipate and forecast, I don't have any real solutions. Even if I had filed a flight plan, the situation and procedure would still have been the same." There is no evidence of the pilot checking enroute for a weather update or a PIREP on conditions. This may very likely have been a solution. The pilot encountered poor weather and simply pushed on. Certainly, a timely retreat would have been a solution. Presumably, the pilot was referring to a VFR flight plan. If that's true, a VFR flight plan would have had no significance, or would it? The pilot reported losing flight following services. Had the airplane gone down, a VFR flight plan would have alerted search and rescue. It appears that an IFR flight plan with weather updates enroute would have been the best solution.

The do's and don'ts of flying in low ceilings and visibilities are

- Do obtain a standard weather briefing.
- Do update weather enroute.
- Do get into the IFR system before encountering poor weather.
- Do ask for help whenever the situation becomes doubtful or uncertain.
- Do have options and a plan for each option.
- Don't let the briefer make the decision (VNR—VFR flight not recommended—or no VNR).
- Don't fly below legal or personal minimums.
- Don't run out of options; retreat or land before you do.

If used, these strategies can prevent low-ceiling and low-visibility accidents. Taking everything into account, all such accidents are preventable. Almost one-third of the pilots involved in low-ceiling accidents and two-thirds involved in low-visibility accidents had no record of a weather briefing. And, although there are no specific records, it's doubtful they obtained weather reports enroute.

4

Turbulence

Turbulence is any disturbance in air flow on a scale small enough to change an aircraft's attitude or flight path. An aircraft experiences turbulence as it flies through sudden changes in the air's direction and speed. Turbulent currents can be either horizontal or vertical and from barely perceptible to extreme in intensity.

Mechanical Turbulence

Mechanical turbulence is caused by any object placed in a moving current of air that impedes the flow. As the current closes in behind the object, eddy currents develop leeward of the obstruction.

CASE STUDY As reported: The approach started with a crosswind entry to runway 14 and a normal left pattern throughout. The wind sock was observed to be almost fully extended but lined up quite well with runway 14. Light turbulence was encountered in the pattern on approach. Flaps were added on final and airspeed was gradually decreased down final. Airspeed was normal for a short runway, about 2100 ft. A normal landing was in progress over the threshold, but with about 10 ft of altitude remaining, some sort of wind shear was encountered and all lift was suddenly lost. The plane dropped and a very hard impact was sustained.

Four impact marks were observed on the runway surface where the propeller struck the runway. While walking around the approach end of the runway immediately after the incident, my personal observation was that the wind was swirling around quite a bit while the windsock indicated a fairly steady wind. In my opinion it is quite possible that a significant wind shear did exist at the time and point of landing and that this is what caused the incident. To the best of my knowledge, there was nothing that could have been done to prevent this incident. The approach was normal in all respects until the very end.

The pilot further stated, "Wind shear is so sudden and unpredictable, I have no recommendations to offer that will prevent a similar event. It is my opinion that the hills, river, and proximity to the ocean can contribute to the wind shear potential at this airport."

Unfortunately, without the weather at the incident site, only general conclusions can be drawn. The runway environment is very important. In this case hills, a river, and ocean can generate significant mechanical turbulence, even with relatively low winds. From the pilot's account it does not appear this was a wind shear event. It does, however, illustrate the significance of mechanical turbulence. The pilot reported the windsock almost fully extended but lined up quite well with the runway.

The pilot elected to use a short field technique because of the length of the runway but did not consider gusty wind conditions, which were implied by the pilot's observations of the weather conditions. Under such a situation of wind and terrain, the windsock may not be representative of conditions at different locations on the airport or even at elevations above that of the sock. Most manufacturers recommend additional airspeed in gusty wind conditions. A pilot making a short field approach is only a few knots above stall. In the flair this margin is further reduced. At this point even a gust factor as small as 5 knots can result in this type of incident. A pilot must balance this against runway length and then determine if the aircraft, under given conditions of wind and runway, is capable of making a safe landing.

How can we determine wind conditions? More and more airports have either towers or automated weather reporting systems. If not, check with a nearby field, or observe smoke or trees on the ground. Note wind drift in the pattern and plan accordingly. Finally, if anything isn't right, go around. The same applies to the takeoff roll; if anything isn't normal, abort.

What are your personal minimums? What is the capability of your aircraft? Exceeding either is a recipe for disaster.

Mountain Waves

Mountain waves occur when wind blows perpendicular to a mountain range with a stable atmosphere. The intensity and significance of turbulence will depend on the wind speed. The more perpendicular to the range and the stronger the wind, the more severe the turbulence.

CASE STUDY As reported: We had climbed to FL310 out of Denver. There were reports of moderate to severe turbulence above FL350. We had just leveled off at FL310 fifty miles west of Denver. We encountered moderate turbulence on climb but smooth air at FL310. At this point we started to notice some extreme mountain wave action. We told the flight attendants to be seated. At this point we were experiencing 1000 fpm [feet per minute] up and down drafts. Then we were hit by severe turbulence. We had to disengage the autopilot to regain aircraft control and we notified ATC.

The reporter went on to state, "We could not have foreseen this encounter since we were the first aircraft of the morning heading west out of Denver. The forecasts were showing the potential, but no PIREPs." Folks, like icing and thunderstorms, if you fly in areas of high potential for turbulence, sooner or later you going to get caught. With all the injuries, and an occasional death, associated with such encounters, I've noticed the airlines have become more conservative about requiring the use of seat belts and telling flight attendants to be seated and strapped in. This is undoubtedly a prudent procedure. Under such circumstances be prepared for a significant turbulence encounter; if it doesn't occur, consider yourself very fortunate.

CASE STUDY As reported: On a flight from Tonopah, Nevada, to Oxnard, California, the pilot encountered severe turbulence in California's Owens Valley. The pilot received a complete briefing that included advisories for moderate turbulence.

I was in contact with ATC during the incident. The aircraft was unable to climb above the turbulence. I judged that the conditions constituted an emergency and requested a clearance through a restricted area. This was initially denied, then approved. I landed at the Inyokern Airport. After an updated briefing, I learned a SIGMET [significant meteorological information] had been issued and decided to discontinue the flight.

California's Owens Valley is notorious for its turbulence. Although we don't know the winds aloft, it appears they indicated moderate or greater turbulence. Under such conditions a pilot might be best advised to fly a longer course to avoid this area. The reporter, in fact, stated, "The west side of the valley, where civilian aircraft are required to fly, is frequently dangerous due to the prevailing winds and mountain wave activity."

The pilot apparently knew the danger. The reporter went on to say, "I would recommend an immediate review of the boundaries of the restricted area, which forces civilian aircraft to operate in a narrow corridor adjacent to the Sierra Nevada mountains and allows for virtually no options to avoid mountain wave activity."

So what can a pilot do? Knowledge is power; the pilot was aware of significant turbulence in this area. As already stated, an alternate route was one option. The pilot apparently never considered reversing course. Once severe conditions, and an emergency situation, were encountered, the pilot confessed to difficulties and was granted clearance through the restricted area. What else could the pilot have done? During the preflight briefing or just after departure, the pilot could have obtained the status of the restricted area. This would have given the pilot the option of flying through the restricted area if it were not in use or deviating to the east to avoid the Owens Valley and the restricted area.

Jet Stream

Clear air turbulence (CAT) associated with the jet stream is very patchy and transitory in nature. The dimensions of these turbulent patches are quite variable generally on the order of 2000 ft in depth, 20 miles in width, and 50 or more miles in length. The patches elongate in the direction of the wind. The dimensions of these areas are on the microscale, and the exact position of specific areas is difficult, if not impossible, to forecast.

CASE STUDIES In December 1998 a passenger aboard a United Air Lines Boeing 747 died as the result of head injuries suffered when the aircraft flew into an area of severe CAT over the Pacific Ocean. The passenger was not wearing a seat belt.

Other less severe injuries have resulted from CAT, again mostly due to the failure to wear seat belts.

In November 1997 I was riding "jump seat" on a Delta Air Lines Boeing 737 from Oakland to Salt Lake City. At FL370 the ride was smooth to occasional light turbulence. Over the radio we could hear a Delta Boeing 757 constantly complain to ATC about the moderate turbulence at FL390. As chance would have it, I flew on that airplane, with that crew, from Salt Lake City to Washington's Dulles Airport. They had flown their first leg from San Jose to Salt Lake City, almost the same route I had flown at FL370 in the 737.

This illustrates how the intensity of jet stream CAT can change significantly over a relatively short distance.

CASE STUDY A Boeing 747 experienced severe turbulence, which caused injuries on a flight from Tokyo to Honolulu. The initial encounter resulted in a gain of 500 ft to FL335 and then a 1500-ft loss to FL320. The temperature dropped from −37 to −40°C in about 2 seconds. The reporter stated that the turbulence was so severe that the instrument could not be read, except in peaks and valleys. The turbulence lasted several minutes. A TCAS [Traffic Alert and Collision Avoidance System] alert was received for an airplane at FL350. All that plane experienced was 100-fpm up and down drafts.

A factor associated with the jet stream is wind shear turbulence. With an average depth of 3000 to 7000 ft, a change in altitude of a few thousand feet will often take the aircraft out of the worst turbulence and strongest winds. Maximum jet stream turbulence tends to occur above the jet core and just below the core on the north side. Additional areas of probable turbulence occur where the polar and subtropical jets merge or diverge.

Moderate or greater CAT occurs most frequently during January and February. Jet stream winds are at lower altitudes, more frequent, and stronger during the winter than in summer.

CAT is greatest in areas of strongest shear. Expect significant turbulence in areas of sharp troughs, the neck of cutoff lows, and in a divergent flow. Severe turbulence can exist despite relatively low wind speeds. Wind speeds and curvature of contours provide a clue to clear air turbulence. Often the curvature of the contours has more effect on the severity of the turbulence than does wind speed.

Wake Turbulence

Wake turbulence encounters with the Boeing 757 have received a lot of press lately. A Boeing 737 captain said a 757 wake rolled his aircraft 45° on approach to Salt Lake City. That must have been exciting. Keep in mind that wake turbulence from much smaller aircraft, including helicopters, can significantly affect small aircraft.

Although wake turbulence from larger aircraft is a factor enroute, it is a more serious concern during takeoffs and landing. During the takeoff and landing phase, aircraft are close to the ground and at relatively slow speeds.

CASE STUDY As reported: After receiving our clearance from ground control we taxied to runway 16. We were number 3 for departure. The aircraft immediately preceding us, a DH8 commuter, was cleared for immediate takeoff. With no wind, I attempted to sidestep possible wake turbulence after rotation but was unsuccessful. The vortex caused our aircraft to make an abrupt roll to the left. Positive recovery was made and my copilot advised the tower that we encountered wake turbulence. No wake turbulence caution was given by the tower, nor was any wait put into effect at that time.

Because of the hazards of wake turbulence, controllers are required to apply specific minimum separation between aircraft of various weight categories. This will be in the form of time or distance. For the purposes of wake turbulence separation minima, ATC classifies aircraft as heavy, large, and small as follows:

- *Heavy.* Aircraft capable of takeoff weights of more than 255,000 pounds whether or not they are operating at this weight during a particular phase of flight.
- *Large.* Aircraft of more than 41,000 pounds, maximum certificated takeoff weight, up to 255,000 pounds.
- *Small.* Aircraft of 41,000 pounds or less maximum certificated takeoff weight.

Pilots must understand that with small aircraft, as defined above, there is no wake turbulence advisory or additional separation required. This, however, does not relieve the pilot of ensuring adequate wake turbulence avoidance. The pilot, especially under conditions favorable to this hazard, should request additional delay or distance for takeoff and landing.

CASE STUDY The flight from Van Nuys to Long Beach occurred on a typical Los Angeles Basin day. Above the haze there was clear, smooth air. I spotted a "three-holer" [Boeing 727] on approach to Los Angeles International. It was above my flight path, so I planned to cross about 3 miles behind. I penetrated the wake at a 90° angle. It grabbed the airplane and almost instantaneously dropped me about 50 feet. Because I was belted in I was also yanked down. My recollection of the incident was seeing my note pad flying out of my pocket in front of my face and two ash trays crossing in front of me, dumping butts all over the place. I remember the first thing I did was check to see if the tail was still there; it was.

The 727 was heavy, clean, and slow. There was little, if any, wind at my altitude—ideal conditions for the generation of wake turbulence. Three miles is not enough separation to prevent a wake turbulence encounter under these conditions.

CASE STUDY As reported: We were being vectored for our climb. ATC was also vectoring a DC10 at FL180 above us. ATC issued a traffic advisory regarding the DC10 at 12 o'clock, FL180 and cleared us to climb to FL230. ATC advised that we were 5.5 miles behind the DC10, "caution wake turbulence." At 16,500 ft we encountered a brief period of turbulence consisting of a rapid shuddering sensation followed by a roll to the left of approximately 10 to 15°. ATC should have offset our flight paths so as not to cross us directly beneath the heavy DC10.

A callback conversation with the reporter revealed that the pilot would not again accept a clearance behind a heavy aircraft with such minimal separation, even though the pilot indicated that he had not encountered wake turbulence from a heavy aircraft before.

ATC is required to separate aircraft operating directly behind, directly behind and less than 1,000 feet below, or following an aircraft conducting an instrument approach by

1. Heavy behind heavy—4 miles.
2. Large/heavy behind B757—4 miles.
3. Small behind B757—5 miles.
4. Small/large behind heavy—5 miles.

Here again, ATC was applying required wake turbulence separation. The reporter remarked that he would not accept such a clearance in the future. This appears to be sound advice. And, let's cut ATC a little slack. They are criticized in one breath for letting aircraft get too close and in the next for delaying aircraft by assigning too much separation.

CASE STUDY As reported: Our Boeing 767 was 10 miles in trail of a Boeing 777 at FL370. The wind was 80 knots on our tail. Wake turbulence created temporary loss of control, an uncommanded roll to the left. We were unable to counter with full right aileron. We reached 45° angle of bank before the turbulence subsided, allowing flight controls to be effective again and recovery.

This case study illustrates the fact that heavy aircraft can produce significant enroute turbulence even with double the minimum required ATC wake turbulence separation. The bottom line is: pilots beware.

CASE STUDIES In one fatal accident ATC placed a Piper Navajo too close behind a Boeing 727. The Piper was less than 3 miles behind the 727 and above its glide path when cleared for the approach. Apparently, the Piper was in a steep rate of descent, trying to intercept the glide slope, when it encountered the 727's wake and crashed.

A Twin Commander pilot following a 727 reported severe turbulence. Although the Commander was nearly 5 miles behind the jet, the NTSB believes the pilot most likely lost control due to slow speed and possible icing and wake turbulence.

Pilots can never allow ATC to put them in an untenable position. The Piper pilot should have initiated a missed approach rather than attempt to dive to the glide slope. Situation awareness was also a factor. The proximity to the jet should have been another clue that a missed approach was in order. Apparently the Commander pilot allowed the airspeed to decay well below normal approach speed. This, coupled with possible ice and wake turbulence, led to disaster.

CASE STUDY As reported: Approach turned us onto a 15-mile final about 2¼ miles behind another MD88, told us to slow to 150 knots, and cleared us for the visual approach. We had the preceding aircraft in sight and thought that, because we were slowed to 150 knots, the spacing would improve. Two miles before the outer marker, we were within 2 miles and closing. We expressed our concern to approach about the spacing, and the controller told us it was fine and handed us off to the tower. We considered asking for an S-turn but could see that the aircraft behind us was within $2^{1}/_{2}$ miles on the TCAS. We told the tower the spacing was too close, and the controller told us to expect landing clearance on short final. We were 600 ft when the aircraft in front of us touched down. As we came over the approach lights at 100 ft, the runway was clear, but we hit wake turbulence. The turbulence required full left aileron to keep the wings level. I immediately initiated a go-around.

The reporter went on to say, "Approach routinely gives us $2^{1}/_{2}$ miles or less spacing on final, assigns us a speed, and hopes it works out." A callback conversation with the reporter revealed that the airlines, pilot groups, and ATC are unwilling to take on the problem. In fact, the reporter was told that spacing was going to go to 2 miles separation, and a few missed approaches were acceptable.

Pilots are certainly "between a rock and a hard place" on this issue. Airline management wants the FAA to reduce separation at many busy terminals. Like icing and thunderstorm, when is enough, enough? If you're not comfortable with the situation, don't accept the clearance or go-around. I know if I were in the back, I would prefer a go-around outside of the outer marker than 100 feet above the runway. Let's hope this issue is not resolved using what some have coined as the FAA's "tombstone mentality."

CASE STUDY As reported: We were in the process of making a visual approach to runway 28L at San Francisco. ATC was setting up simultaneous visuals to both 28L and 28R. We had previously reported a Boeing 777 in sight on visual to 28R. On approximately a $1/_{4}$-mile final at 400 ft we experienced turbulence and uncommanded roll. We initiated a go-around. I noted the FMS [flight management system] computed winds of 035° at 10 knots. Our second visual approach and landing were normal.

The reporter suggested that ATC make "winds aloft" requests from the numerous FMS-equipped aircraft in the area. ATC already has a significant wake turbulence burden. However, this incident does suggest that aircraft equipped with FMS, or other means of determining winds at altitude, can be used as another source to determine the probability of a wake turbulence encounter. Additionally, it appears that standard wake turbulence avoidance technique would have resolved this incident—remaining above the other aircraft flight path and landing beyond the touchdown point of the heavier aircraft. This assumes, however, that the reporting aircraft had sufficient runway to accomplish this maneuver.

CASE STUDY As reported: On approach 20 miles out we were told to slow, that we would be following a heavy aircraft. We complied, were descended and told to capture the localizer. We were then cleared for the approach. We were 6 miles in trail of a heavy. Weather conditions at the time were broken ceiling at 800 ft, visibility 6 miles. We were switched to the tower and cleared to land. We could see the preceding traffic at all times, and the runway intermittently. At about 1500 ft, we saw the heavy in front of us start a go-around. We hit its wake at about 1200 ft as we were entering the cloud tops. It rolled our aircraft 20° in either direction as we passed through its wake.

The tower never said a word about the preceding aircraft making a missed approach. After landing, I called the tower. The supervisor told me that the Boeing 747 was unable to maintain the spacing required on the Boeing 757 in front of it and was instructed by ATC to go around.

When I asked why were weren't told of the go-around, the supervisor said that the controllers were "too busy" with the spacing issue to issue any further advisories.

The reporter placed the blame for this incident on ATC and the lack of information contained in the *Aeronautical Information Manual* (*AIM*). The reporter stated, "The *AIM* devotes 7 pages to wake turbulence. Nowhere in there does it reference this situation. If there is any situation guaranteed to establish a wake turbulence encounter, this would be it. . . ."

Unfortunately, the report does not state if both aircraft were on the same frequency. Had they been, the reporting crew should have been aware that ATC had instructed the 747 to go around. This is "situational awareness." ATC was involved in a separation issue, its primary task; and as the reporter discovered, ATC has no requirement to advise the trailing aircraft in such a situation.

CASE STUDY As reported: We were being vectored crosswind, 15 miles from the runway. The controller assigned 160 knots to follow a Boeing 757. I was told to "advise if unable," and I refused the speed for wake turbulence considerations. The controller replied with an expected delay vector across final for spacing. Shortly we were turned to final and almost immediately encountered wake turbulence. The autopilot disengaged and the aircraft flew off final. TCAS showed the preceding aircraft $4\frac{1}{2}$ miles ahead and 1100 ft lower. A high final was flown and where we were handed off to the tower, we requested the parallel runway. The tower was unable to approve the request.

After landing, I asked the tower why they were not able to approve the request for the other runway. They responded an aircraft was cleared to cross that runway. I explained we had requested it for wake turbulence considerations and had encountered wake turbulence on final. Each controller had stated ". . . following heavy, use caution . . ." but then slowed me down and refused me an alternate runway. It would appear more consideration could have been applied to avoid a possible dangerous situation.

This is good example of the pilot exercising pilot-in-command authority. The pilot refused the speed reduction for wake turbulence considerations. The pilot then applied sound wake turbulence avoidance techniques. Had the pilot requested the alternate runway earlier, it might have been approved. Pilots must understand that if the runway has been given to another controller (ground control), depending on the situation, it normally cannot be immediately revoked.

CASE STUDY As reported: While attempting to intercept the localizer from the north, we encountered wake turbulence from the preceding heavy aircraft. The encounter was severe enough to raise concern over the safety of the aircraft; therefore, a full prescribed recovery was initiated. However, in the attempt to maintain control of the aircraft, we flew through the localizer, with this putting us more in line with approaches to the parallel runway. ATC advised us to fly a 330° heading to reintercept. We advised approach of the event and proceeded with a normal approach and landing.

The use of rudder as demonstrated in our training, I really feel, made a big difference in our roll recovery rate.

During a callback conversation the reporter stated that he just completed the unusual attitude training and believes that helped him to recover with minimal problems. He feels that ATC could help by bringing the heavier aircraft in at a lower altitude than the lighter aircraft. He feels the quartering tailwind caused the wake to remain in his path.

Interestingly, the FAA has relatively recently recognized the hazards of a quartering tailwind. A light quartering tailwind can move the vortices of the preceding aircraft forward into the touchdown zone and requires maximum caution. This is now a question on the FAA's written exams.

CASE STUDY As reported: While conducting currency night landing, I encountered the rotor wash from two H53 helicopters on short final, resulting in a go-around. I was downwind when a military helicopter called, entering the pattern. I did not see the traffic and asked for its position. They replied a flight of two on final. I then picked them up visually. All I saw was two steady red lights about 100 ft apart. I did not see a flashing beacon or strobe. As I approach abeam the end of the runway, the helicopters were on short final. Because I figured they would be on the ground for a while, I extended downwind, called turning base and final. On short final, at about 100 ft, I experienced severe turbulence. The Cessna 150 went into an uncommanded 30° bank with severe buffeting. I immediately did a go-around. On the next pass, I extended 2 miles upwind while the H53s made another landing.

The reporter commented that the military helicopters did not use proper uncontrolled airport entry, pattern, or voice procedures. Unfortunately, these procedures are recommendations rather than requirements. The reporter believes they are required to use the fixed-wing pattern and comply with fixed-wing procedure. Helicopters are typically required to avoid the fixed-wing pattern. The reporter stated that he was unaware of the size of the helicopters and would have left much greater separation had he been aware. The H53s did not announce they were "H53s" or "heavy." He feels strongly that it is important at night for traffic to announce position and type. This pilot could contact the FAA's Aviation Safety Specialist and relay his concerns to the military. I have under similar circumstances, with positive results. The military is as concerned with safety as we are.

CASE STUDY As reported: While being vectored for an approach, we experienced what I believe was severe wake turbulence. We were on an assigned heading at 11,000 ft and assigned speed of 250 knots. It started with maybe 5 seconds of steady but very light turbulence. I commented on the strange feel to the other pilot. As I said that, the airplane rolled into a 90° bank to the right instantly. I disengaged the autopilot and recovered by hand-flying the airplane. By the time the wings were level, we had lost 700 ft of altitude. We told ATC what had happened. The controller said that we were 8 miles in trail of a Boeing 757.

I called both the tower and approach control after landing and talked further with them. Given the speed of the 757 and our airplane, the controllers were surprised that the wake turbulence was so severe.

A callback conversation with the reporter revealed the following. The reporter states that his review with ATC brought added questions because the 757 was in the clear and at 250 knots. Both aircraft were on radar vectors and on the exact same flight path and speeds. The in-trail separation should have been ample to avoid such a wake turbulence encounter. The reporter was surprised when the autopilot was disengaged that there was not control input already active. In discussion with the manufacturer, the reporter learned that in extreme control changes the autopilot goes into a "coast mode" until it can determine if the extreme is a false input or not. The autopilot then responds with correct input. The reporter says, "you learn something new all the time."

Here are some final comments about the requirements of ATC. ATC must separate an aircraft landing behind another aircraft on the same runway or one making a touch-and-go, stop-and-go, or low approach by ensuring the following minima will exist at the time the preceding aircraft is over the landing threshold:

(Parallel runways less than 2500 feet apart are considered as a single runway because of the possible effects of wake turbulence):

1. Small behind large—4 miles
2. Small behind B757—5 miles
3. Small behind heavy—6 miles

Note: In some terminal areas, under certain conditions, a separation of 2.5 nautical miles is authorized between aircraft established on the final approach course within 10 nautical miles of the landing runway.

Here's the disclaimer:

> Vortex avoidance during VFR operations must be exercised by the pilot. Wake turbulence may be encountered by aircraft in flight as well as when operating on the airport movement area. Pilots are reminded that in operations conducted behind all aircraft, acceptance of instructions from ATC . . . is an acknowledgment that the pilot will ensure safe takeoff and landing intervals and accepts the responsibility for providing wake turbulence separation.

Remember, pilots can request additional wake turbulence separation. Like icing and thunderstorms, avoidance is the key to the wake turbulence hazard.

5

Icing

Icing is as difficult to forecast as it is hazardous. Forecasters must determine which areas contain enough moisture to form clouds, which cloud areas will most likely contain supercooled droplets during the forecast period, and the freezing level. Needless to say, this is not an easy task. Pilots should consider these icing advisories as forecasts of icing potential. They alert the pilot to the need to consider the possibility of icing in clouds and precipitation within the areas and altitudes specified.

For most light aircraft, avoidance is the only solution. Once flight instruments have failed, the engine has faltered due to carburetor ice, or the airfoils incapacitated by ice, the pilot has few, if any, options.

Ground Icing

Ground icing is a hazard produced by snow, ice, or frost on the aircraft or runway. As we shall see, ground icing can be extremely dangerous.

CASE STUDY Some years ago four of us flew a Cessna 172 into the South Lake Tahoe Airport in November. After our stay, about three-quarters of an inch of ice had formed on the airplane. Being naive at the time about such conditions, I assumed it would blow off during the takeoff roll—silly me.

An experienced tower controller, however, suggested we clean the ice. It had to be scraped off with a plastic scraper!

Later I calculated we had between 300 and 400 pounds of ice on the airplane. I had no experience with this condition at the time. I shudder to think what would have happened during a takeoff in a low-performance airplane, 300 pounds over gross, at a high-altitude airport, with three-quarters of an inch of ice on the airfoils. This is a perfect example of where the test almost came before the lesson.

CASE STUDY A Cessna Caravan pilot was parked outside in freezing rain. The next day the pilot removed about 80 percent of the snow covering the wings, which left a coarse layer of ice about three-sixteenths of an inch thick. The aircraft crashed after takeoff. The probable cause was determined to be the pilot's failure to remove ice from the airframe prior to takeoff.

It's foolish to attempt to take off with any snow, ice, or frost on the airplane. Snow, snow melt, freezing rain, and frost produce structural ice that can be difficult to remove from a parked aircraft. Even a thin layer of frost can severely affect performance and must be removed before takeoff, according to federal regulations.

The solution to these problems is to hangar the aircraft, arrange for a deicing service, or plan the departure later in the day when the sun has melted the ice. During taxi, avoid areas of standing water or slush. Slush thrown into wheel wells, wheel pants, and control surfaces can freeze, resulting in locked controls and frozen landing gear or brakes. With a descent through an icing layer, and surface temperatures close to or below freezing, a pilot must be prepared for an approach and landing with airframe icing, possibly on an ice- or snow-covered runway.

CASE STUDY The investigation of a Cessna 152 accident revealed that the crankcase breather line was plugged with ice, and the oil had been forced out through the engine's nose seal. The probable cause was determined to be the pilot's inadequate preflight check that failed to detect the ice-clogged breather.

If an aircraft is parked in an area of blowing snow, special attention should be given to openings in the aircraft where snow can enter, freeze solid, and obstruct operations. These openings should be free of snow and ice before flight.

CASE STUDY The low-time pilot said he could see runway lights prior to the nighttime takeoff of the plowed runway. The pilot aligned the airplane for takeoff between the snow banks along the sides of the runway and by sighting down the runway. The takeoff roll was normal until rotation. As the airplane lifted off, it struck the snow bank on the right side of the runway and nosed over.

A pilot's first source of surface conditions are Notices to Airmen (NOTAM). A check with the FSS or Direct User Access Terminal (DUAT) should reveal airport conditions. However, under conditions of snow or ice, it's often a good idea to inspect the takeoff area just prior to departure. This is especially true at night and for airports without an operating control tower. This procedure will allow the pilot to determine any hazards or anomalies on or near the runway environment.

CASE STUDY As reported: On landing rollout at approximately 1000 ft down the runway we encountered a slick spot on the runway. With the crosswind the airplane slid sideways off the runway. The airplane nosewheel went into a snowbank and both wingtips were struck.

A callback conversation with the reporter revealed the following: The reporter said that he did not ask, and the UNICOM operator did not volunteer, runway conditions during the approach. He knew about the crosswind, because of his corrective actions during the approach. After touchdown, the aircraft was tracking down the centerline of the runway until it rolled onto a patch of ice and compacted snow. At that point, the aircraft started to weather vane and slide toward the left side and then off the runway. Here again, a pilot's first source of airport conditions are NOTAMs. Additionally, overflying and inspecting the runway may reveal hazardous or unsafe conditions.

If snow, ice, or slush are on the runway, aircraft control might be difficult, especially in high winds, with reduced braking efficiency, resulting in a longer-than-normal ground roll. Regulations require the pilot to consider ". . . runway lengths at airports of intended use. . . ." Pilots operating in this environment must consider these factors when selecting destination and alternate airports or even deciding the advisability of making the flight.

Instrument Icing

Instrument icing is any icing situation that causes loss of, or erroneous, flight or engine instrument indications. Most often the pitot-static instruments are affected.

CASE STUDY As a new instrument instructor I took an instrument student on a flight from Van Nuys to Lancaster's Fox Field. The freezing level was forecast to be at 6000 ft. The minimum altitude for the route was 7000. Cloud bases were at 6500, well above terrain. Sure enough, we picked up trace to light rime icing. We had neglected to turn on the pitot heat and a reverse cone of rime ice grew from the pitot tube. I, matter-of-factly, pointed this out to my student and we turned on the heat, which immediately corrected the problem. On descent, the ice made a deafening noise as it broke off the tail surfaces. I do not recommend this procedure.

The majority of small general aviation aircraft have no, or limited, deice and anti-ice equipment. That which is available typically consists of pitot and carburetor heat and an alternate air source for the pitot-static instruments. Any time a flight is conducted near the freezing level (as a rule of thumb, within 2000 ft) the pilot must preflight any deice or anti-ice equipment onboard the aircraft and ensure its proper operation. Some manufactures recommend the use of pitot heat any time the temperature is less than 40°F (4°C) in visible moisture.

CASE STUDIES A Boeing 727 was lost due to an iced-over pitot tube. As static pressure decreased during the climb, the airspeed indicator showed speed increasing. The autopilot attempted to hold airspeed by increasing pitch, resulting in a stall. A Boeing 737 crashed because of an iced-over engine power sensor—the airplane was simply not developing takeoff power, even though the instrument indicated so.

The solution to an iced-over pitot or engine instrument probe is crosscheck. Crosschecking all flight or engine instruments will reveal the discrepancy. With the problem diagnosed, a plan can be developed to work around the situation.

Induction Icing

The induction system includes the air filter, ducting, and fuel metering device. Induction system icing consists of any ice accumulation that blocks any component of the system. The majority of general aviation icing accidents are attributed to carburetor or induction system icing.

CASE STUDY On a flight from Van Nuys, California, to San Francisco in a Cessna 172 we encountered light icing after an ATC instruction to climb. I periodically applied carburetor heat. Something unusual occurred. With carburetor heat on, the engine ran fine; off, the engine faltered. On the ramp at San Francisco we parked next to a Navion that had also flown from Los Angeles Basin, but at a higher altitude encountering more ice. Sure enough, in the Navion's air filter was a large chunk of ice. I realized that the carburetor heat in the 172 was functioning as an alternate air source. I'm sure this seems ridiculously obvious; it didn't at the time, which illustrates the hazards of learning by experience.

In addition to air intake icing, normally aspirated engines can develop ice in the carburetor throat.

CASE STUDY I had remained overnight in Amarillo, Texas, because of a line of thunderstorms that approached from the west. The Cessna 150 was parked into the wind when torrential rains moved through the area. The next morning was clear, with abundant surface moisture, temperature was about 15°C, and nearly 100 percent relative humidity.

My first clue of trouble was the increased throttle setting required to obtain idle rpm. Engine runup also took more throttle than usual. I suspected carburetor ice and a water-saturated air filter because of the conditions. I had a 13,000-foot runway.

Full throttle only gave me about 2200 rpm. The increased ground run to rotation speed—about 7000 ft—should have been another clue. I was off the ground and with no runway remaining and 200 feet of altitude the engine started losing rpm. I applied carburetor heat and the engine was running very rough, producing about 1700 rpm.

There was a tremendous psychological urge to reduce carburetor heat and get that rpm back. I was preparing to crash straight ahead, but the engine was still producing power and I decided to make a 180° turn and land on a taxiway. Then I informed a surprised tower controller of what happened; remember, a pilot's first job is to fly the airplane.

After running the engine for 20 minutes, and one aborted takeoff later, I launched into the air. The engine again performed normally above the shallow, moist layer. It was a perfect example of having the clues and ignoring them. I was extremely fortunate.

Accepting a Clearance

A pilot's acceptance of an ATC clearance is a contract. It obligates ATC to provide separation between the pilot's aircraft and other known aircraft within the system; it obligates the pilot to comply with all aspects of the clearance, except when the pilot declares an emergency. These obligations are clearly spelled out in the regulations.

CASE STUDY As reported: On the previous leg we noticed sparks coming from the left engine. A closer inspection revealed that the engine anti-ice protection had burned up. Because the weather was currently VFR at our destination and forecast to stay that way, we both decided that it would be OK to do the next leg where maintenance could then work on it. The captain told me that he called maintenance about the problem. Because we told them about it and wrote it up in the logbook, I thought all was well. The problem was that even though it was VFR, we still accepted the IFR clearance, not anticipating that any clouds would be on our route. We were cleared from 8000 to 6000 ft enroute where we encountered a scattered layer. I asked ATC if we could get lower as soon as possible. ATC said that it would be soon. My captain then told me to explain that we needed lower because we could not go into any clouds. I said that if we tell ATC we can't go into clouds on an IFR flight plan, we should declare an emergency. The captain said that because we wrote up the problem in the logbook that it would be all right. So, I told ATC the problem and within a few miles we were cleared to 4000 ft and in the clear.

The reporter went on to say, "Looking back on it, we should have gone VFR because we had no ice protection. It was only a scattered deck at 6000 ft, but the captain was afraid we might enter a cloud which might flame out the left engine. Being on an IFR flight was a mistake because that implies that we can penetrate the clouds if we have to, which we could not. I felt comfortable with making the flight under the weather conditions. However, I should have suggested we go VFR. Another contributing factor was the relationship between the captain and myself. The captain is a bit of a know-it-all, overbearing, and rarely asks for my opinion. This was one of those occasions where more discussion probably would have prevented more problems."

What other options were available to the crew? When it became apparent the aircraft would enter clouds, the crew could have requested a deviation, to avoid the layer. If ATC can approve a deviation, they will. However, approval is based on traffic, and a pilot can never absolutely count on this procedure. If it became apparent clouds would be entered, the crew could have simply canceled IFR. From the report it appears there was no problem maintaining VFR. The bottom line, however, as stated by the reporter is that the crew should never have accepted the clearance with a known equipment deficiency.

Icing

CASE STUDY As reported: I obtained a standard weather briefing. Any icing at 6000 ft would be north of my planned route. I filed for 4000 ft, but 6000 ft was the MEA [Minimum Enroute Altitude] in my direction of flight. Believing that any icing was north, I accepted the clearance of 6000 ft. Cruising at 6000 ft, I noted the beginning of ice on the leading edges of the wings and none on the windscreen, prop tips, or leading edges of the stabilizer that I could detect. I informed ATC of the situation and was reminded of the MEA. The accumulation was trace to light. ATC gave me a clearance to 5000 ft and the ice melted at 5800 ft.

The reporter went on to say, "Weather is not an exact science, and we need to realize this fact and use our judgment accordingly!" The reporter is quite correct. The reporter knew the MEA was 6000 ft. Keep in mind that trace icing was not forecast. However, as pilots we should always expect some icing in visible moisture above the freezing level. From the case study it appears the pilot took positive action to get out of the icing as soon as it was detected, which was the correct procedure.

CASE STUDY As reported: During a long IFR cross-county training flight, we were in IMC near Reno at 12,000 ft when we encountered light mixed ice. I requested a lower altitude and was advised it would be about 5 minutes due to other traffic. During the wait we lost our airspeed indicator—pitot heat was on. I informed ATC of the situation and became adamant in my request for a lower altitude. ATC issued a descent to 11,000 ft. My student began a hasty descent, due to the ice situation. During this time the static system began to fail. I instructed my student to select alternate air, which he did, and the static system was restored. At that point we realized that the indicated altitude was 10,400 ft. After we arrested the descent, ATC asked our altitude, I replied 10,400. I mistakenly reported that I understood the assignment was 10,000 ft. Due to the tense situation, in want of an excuse I wrongly stated this, knowing that the assignment was 11,000 ft.

We checked PIREPs on ice. We were certain that our trip would keep us out of the ice. We diverted to Reno as a precaution and stayed the night to continue our trip the following day.

The reporter states PIREPs on ice were checked and that they ". . . were certain that our trip would keep us out of ice." No mention was made of checking icing forecasts. Recall from the last case study: Always expect some icing in visible moisture above the freezing level.

There is a myth in aviation that only PIREPs constitute "known icing." Let's put that myth to rest. A NTSB decision in 1993 held a pilot in violation of Federal Aviation Regulations. The Board found that a pilot cannot pick and choose between forecasts and PIREPs. A forecast for icing is sufficient to warrant the violation.

Loss of the airspeed indicator makes the aircraft unairworthy. By regulations this was an emergency situation. Remember ATC's functions. If you need to get ATC's attention immediately, declare an emergency! However, there is no "free lunch"; you may have to explain the situation. There are two things to keep in mind. First, you always want to be able to represent yourself at the NTSB hearing. Second, more often than not ATC is so busy and so happy to get us out of their hair, we'll never hear another word. No one should interpret this as meaning that I, the FAA, or ATC condone such actions. Icing for aircraft not certified for flight in icing conditions is to be avoided. At present, the decision as to whether the flight can be made safely rests solely with the pilot, which is where it will stay until we prove we are not worthy of the responsibility.

Let's make another point about responsibility. This pilot showed lack of pilot-in-command responsibility in two ways: allowing the airplane to get into an icing situation in the first place and then fibbing about the altitude. Everyone has busted an altitude. ATC doesn't care unless separation is lost. If it has, fibbing isn't going to do you any good. Everyone knows, or should know, that all ATC communications are recorded. What I teach, or maybe it's preach, is to advise ATC immediately of any significant altitude deviation. This at least allows them to do their job—get anyone else out of the way. If loss of separation does not occur, no big deal; if loss of separation has occurred, they're going to get you anyway. We all have to take responsibility for our actions.

I've been there and done that. Pilots have to understand that there are times when ATC cannot assign the requested altitude. Like the pilot in this case study, I was assigned an altitude that put me into icing. I informed the controller, who replied, "In that case you can declare an emergency or go into Santa Barbara." ATC will, however, do their best to get you your requested altitude as soon as possible. In my case I specifically asked when I could expect lower. The controller said in about 15 miles. Like this pilot, the icing was not significant for the short exposure, but I had a plan. Once ice is encountered, immediately develop a plan to get out of the icing environment.

Enroute Icing

For our purposes enroute icing is an ice encounter during climb, enroute, or on approach.

CASE STUDY As reported: On the ILS approach we encountered unexpected severe icing, which subsequently caused the aircraft to depart controlled flight. The stall/departure indications were consistent with a "tailplane stall." This resulted in an uncontrolled descent below the glide slope while the crew worked diligently to recover the aircraft. The crew recovered controlled level flight at 1600 ft and maintained level flight above the localizer MDA [Minimum Descent Altititude] until glide slope interception. Prior to glide slope intercept, the crew performed a nominal control check and flew a normal landing at Vfe 30, 150 KIAS.

As the flaps extended to 30°, the aircraft displayed a dramatically increasing crescendo of high-frequency control flutter and low-frequency airframe buffeting. As the flaps reached 30°, the aircraft decelerated abruptly and departed controlled flight. At departure, the aircraft rolled left wing down and pitched down 4°. The estimated 15°-per-second roll was uncommanded and opposite to the 30-to-40° right control wheel input. This roll was arrested at 10 to 12° left wing down by positive forward stick input. The crew recovered the aircraft by applying basic airmanship: simultaneously reselecting flaps 15°, increasing power, and applying positive forward stick. Concerned that the roll may be torque related, the crew applied maximum power. The captain maintained right roll control wheel input from the onset until the wings responded, and the wings matched the wheel so that the captain leveled the aircraft very gradually. Initially the aircraft required 100 percent power for level flight, but as the ice protection system functioned, the captain was able to retard power. Throughout the recover, the captain's control inputs were smooth, steady, and stable, whereas the aircraft response was slow and very sluggish, especially around the roll axis, with the pitch slowly oscillating until level flight was achieved.

The encounter with severe icing was of an extremely brief duration—estimated less than 1 second—and the crew had no prior airframe indications (i.e., significant side windshield spattering or prop spinner accumulations growing aft).

The reporter indicated there were no warnings of severe icing (Automatic Terminal Information Service,

Icing 77

or ATIS; radar; or PIREPs). Warnings of severe icing are advertised in SIGMETs or Center Weather Advisories (CWAs). SIGMETs and CWAs affecting the terminal area are required to be put on the ATIS. Therefore, apparently there were no SIGMETs or CWAs in affect. Often SIGMETs for severe icing will not be issued until a PIREP indicates these conditions exist. Why? A report or forecast of severe conditions precludes certain operations. Unfortunately, this puts the pilot and forecaster between the "regulations and a hard place." AIRMET ZULU contains warning of moderate icing and may forecast the existence of supercooled large drops (SLDs). A forecast of MXD or CLR icing implies SLDs. A forecast of SLDs means there is a possibility of significant ice forming aft of the ice-protected areas. AIRMETs are not normally carried on the ATIS.

The reporter went on to say that the aircraft's ice protection systems had been continuously operated during flight in icing conditions. The crew had experienced no problems with icing prior to the unexpected encounter with severe icing. The crew briefly observed moderate icing at 3000 ft. Prior to the approach the aircraft had been in icing conditions for approximately 15 to 25 minutes and reported an accumulation of a total of 1 in of mixed rime. There is no such thing as mixed rime. Did the reporter mean "mixed rime and clear?" If that was indeed the case, mixed icing is a clue to possible SLDs.

A callback conversation with the reporter revealed that at time of the incident the side window had suddenly gone opaque from ice accumulation. They concluded that the flutter was a result of icing on the elevator. As flaps were raised to 15°, the flutter stopped. This was reinforced to the crew during the upset training given by their company. The crew missed the first approach and flew for 15 to 20 minutes in IMC icing conditions before beginning the second approach. Conditions were right for picking up elevator icing.

This incident reveals excellent Cockpit Resource Management (CRM) and training. The crew did the right thing at the right time. Icing training videos and training CD modules have been produced by NASA and are available through commercial vendors.

CASE STUDY A nonturbocharged Baron, without ice protection equipment, departed Reno, Nevada, for southern California. Moderate icing and severe turbulence were forecast. The pilot elected to fly a direct course along the crest of the Sierra Nevada mountains, the route where the most intense icing and turbulence could be expected. The aircraft iced up, resulting in a fatal accident.

The pilot had no way out because the MEA was the aircraft service ceiling. The terrain was well above the freezing level and the pilot failed to reverse course at the first sign of ice. What options were available? The pilot could have crossed the mountains near Sacramento, minimizing exposure to ice, and once over the Sierras, it was, literally, all downhill. The pilot could have flown toward Las Vegas where the weather was considerably better or simply waited for better weather conditions.

When the Baron became ice covered, the pilot had no option but to ride it to the crash site. Attempting flight under these conditions and with this type of equipment is quite literally suicide.

CASE STUDY A Bonanza pilot departed the San Francisco Bay area on a flight to Los Angeles. Icing above 7000 ft was forecast and reported. The pilot elected to fly at 11,000 ft. The pilot's last words were, "I've iced up and stalled." The crash occurred in the San Joaquin Valley where the elevation was near sea level. Minimum altitudes in the vicinity of the crash were well below the freezing level.

Especially for aircraft with limited or no anti-ice or deice equipment, the pilot must take immediate action should ice be encountered. The Bonanza pilot had above-freezing temperatures well above the MEA. The pilot simply did nothing until aircraft control was lost.

CASE STUDIES The pilot of a twin-engine airplane was enroute when the airplane encountered icing conditions. When the pilot activated the deice boots, the right wing deice boot failed to function. As the airplane slowed for landing, the asymmetrical ice buildup caused instability that the pilot was unable to control. The airplane crashed on landing, seriously injuring the occupants. There was no mention of the pilot checking the boots prior to departure.

One of our local pilots at Fresno had a similar occurrence in a Piper Aerostar. During the first storm of the season, upon activating the deicing boots, one inflated and the other did not. Our pilot was visibly shaken by the experience.

The bottom line: Check all aircraft equipment prior to departure. Most aircraft used in commercial operations have minimum equipment lists (MELs). Certainly any malfunction of required anti-ice or deice equipment would preclude flight into icing conditions. Those smaller general aviation airplanes that are certified for flight into icing have specific preflight procedures in the approved *Airplane Flight Manual*, which must be accomplished. Know your aircraft, its equipment, preflight and operating procedures, and limitations.

CASE STUDY As reported: A cold front was forecast to move through Arizona and Colorado during our flight. The first leg of this scheduled trip was completely free of inclement weather enroute. Our flight was on the ground for about 10 minutes, and we took another 25 minutes to get about halfway back. Therefore, we first passed this point only about 1 hour earlier. We first encountered what seemed to be light to moderate mixed icing at FL220. Our aircraft is approved for flight into known icing. However, after a short time the heated windshield began icing up, as well as the engine inlet lip area! We noted the outside air temperature of 0°C. We were cleared down to 15,000 ft. The temperature and icing continued all the way down to 11,000 ft. Moderate turbulence occurred throughout the flight, which lasted 2 hours and 25 minutes. A postflight inspection of the aircraft revealed large pits in the fuselage paint adjacent to the props and prop paint nicked off from the ice.

The reporter commented, "All the literature on freezing rain is true—it's dangerous! The B1900D is a fantastic aircraft in icing conditions. However, not even this aircraft comes anywhere close to being able to deal with freezing rain. My best advice: Be aware of and avoid freezing rain at all costs!"

The fact that the pilot continued in the icing environment during the descent indicates an inversion that is indicative of freezing rain. Freezing precipitation often produces severe icing. And no aircraft is tested or certified for flight in severe icing.

CASE STUDY As reported: After takeoff our flight was climbing to 15,000 ft. We began picking up moderate ice at 7600 ft. Ice accumulation continued through 11,000 ft. We requested descent to 9000 ft for ice and it was approved. After a short time at 9000 ft, the situation began to worsen. The aircraft deice and anti-ice systems were functioning properly, but because of "nonshedable" ice on the airframe, we began losing airspeed at the rate of approximately 1 knot every 10 seconds. Because of very heavy traffic on the frequency we were unable to contact ATC to request a lower altitude right away. After approximately 90 seconds we got an opening on the frequency and requested an immediate descent to 7000 ft because of icing. There was no reply. Airspeed now at 160 knots, down from 200 KIAS. I forcefully transmitted that we were "vacating 9000 for 7000 because of heavy icing." The controller responded, "Well, if you're going to do that, turn right to 220°, you have traffic at 11 o'clock at 8 miles at 8000 ft."

The reporter commented, "The problem was extremely heavy radio frequency traffic hampering effective air-to-ground communication and an indifferent attitude of the controller."

Here again, without knowing the weather it's difficult to know if severe icing was reported or forecast. The reporter states that the deice and anti-ice systems were functioning properly and that the ice on the airframe was "nonshedable." This means that the equipment was not operating properly—either mechanically or because of improper operation by the pilot—or that the icing intensity was in fact severe. In either case the pilot should have immediately declared an emergency. This would have allowed the controller to give the pilot proper attention and clearance. Had loss of separation occurred, this individual would have been a candidate for a violation, regardless of the circumstances because the clearance was not adhered to and an emergency was not declared.

CASE STUDY As reported: We were IFR on top at FL190 for most of the flight. The airplane is certified for flight into known icing. All systems were operable. Prior to cloud penetration, following descent instructions from ATC, all deice system were engaged. This included pitot heat, stall warning heat, prop deice, and electric glass windshield on low.

Light rime/mixed icing was encountered on descent. Fifteen to twenty miles west of the airport we began slowing the airplane in preparation for the approach. We continued checking leading edges of wings and spinners—no significant accumulation was noted. The procedure for wing and tail deicing is to wait until there is at least ¼ inch of ice before inflating the boots. Earlier application of the boots could just lift a thin layer of ice, upon which more ice might form, rendering the boots ineffective. Level at 3000 ft at about 140 KIAS the airplane began to buffet. Elevator response became mushy and it appeared the airplane was ready to stall. While I was reaching to add more power, the airplane nosed over and began a left turn. I went with the turn, trading altitude for airspeed while bringing the props and manifold pressure to the top of the green arc and cycling the boots. The airplane felt as if it were operating at the edge of a stall.

At this time I declared an emergency. ATC asked me what I wanted to do. I did not want to put the airplane into any configuration which would result in slower airspeed because I assumed the airplane must have been carrying more ice than I was aware of or there was another problem with the tail. I wanted to do everything I could to avoid another excursion. With the increased power and resultant airspeed and cycling the boots, I got the airplane stabilized and was able to maintain 3000 ft. I climbed to 3500 ft and requested vector for the ILS approach to a closer airport.

My reasoning was that there were less turns involved, lower approach minimums, and a longer runway. I elected to land straight in, although the winds were favoring another runway. I felt comfortable with the tailwind given the runway length, and wanted to avoid turns that would be associated with circling procedures or the additional flying time and higher minimums associated with the other approach. ATC informed me that a Baron had just landed and reported ceiling ragged at 700 ft. We experienced no further difficulty and came to a full stop using only slightly more than half of the runway. During the approach the boots were clear of ice and inspection of the tail after landing showed boots clear there as well. There was some remaining ice on the nose, spinners, upper portion of the tail, and other unprotected areas—perhaps ¼ inch. I assume significant ice was shed while descending.

Perhaps I was too hasty to declare an emergency, but when the aircraft first broke from controlled flight, I had but one single mission on my mind—get the airplane safely on the ground using all the facilities available. At the time I declared the emergency, I wasn't sure whether or not I was going to be able to regain control of the airplane. As it turns out, the extra airspeed and continuous cycling of the boots did the job.

From a more critical perspective, I have now become an even more cautious flight planner. When icing is possible, I will allow for higher ceilings and a cushion of "above freezing altitude" below the bases at the destination airport. I now know first hand that icing conditions are unpredictable and how localized severe icing can be—it can quickly overpower a "known icing" aircraft. We most likely experienced moderate-to-severe clear icing in addition to the mixed/rime, which was not as noticeable from the pilot's viewpoint. In the future I will be quicker to avoid and will expect much worse than "light-to-moderate" forecasts because I realize how quickly things can change for the worse. I will also maintain a higher-than-normal airspeed during flight in conditions that are conducive to icing of this nature. In hindsight, I believe the boots could have been cycled earlier. I am very grateful to ATC for their assistance.

This case study is an excellent example of a pilot taking control of the situation. Notice that the pilot flew the airplane, declared an emergency—which it certainly was—and then used sound aeronautical decisions to divert to an appropriate alternate.

6

Thunderstorms

Thunderstorms contain just about every weather hazard known to aviation—turbulence, icing, precipitation (including hail), lightning, tornadoes, gusty surface winds, low-level wind shear (LLWS), effects on the altimeter, and low ceilings and visibilities.

Convective Low-Level Wind Shear

CASE STUDY In August 1985 a Delta Air Lines L-1011 crashed at the Dallas/Fort Worth Airport. The NTSB was unable to determine if the crew had been using airborne weather radar at the time of the crash. The NTSB report did state, however, "The evidence concerning the use of the airborne weather radar at close range was contradictory. Testimony was offered that the airborne weather radar was not useful at low altitudes and in close proximity to a weather cell . . . ;" although, "at least three airplanes scanned the storm at very close range near the time of the accident."

The accident was probably caused by a microburst from a single severe storm cell, which illustrates how weather can develop rapidly, often without any severe weather warning.

Pilots should avoid takeoff or landing when a thunderstorm is within 10 to 20 miles of the airport. This is the region of the strongest and most variable winds. Caution must also be exercised following thunderstorm passage. A strong, gusty outflow boundary can follow the storm. Along with these winds are downbursts and microbursts that produce severe low-level wind shear.

CASE STUDY American Airlines flight 1420 crashed while attempting to land at Little Rock's National Airport. The crash occurred just before midnight on June 1, 1999. On the plane's approach, a microburst hit the airport with a wind gust to 76 knots. KLIT 020458Z 29010G76KT 210V030 1/2SM + TSGSRA . . . .

Note that in addition to the gust, wind was variable between 210° and 030° to 180°, a classic indication of a microburst.

The FAA, along with a group of aviation specialists, has developed AC 00-54 *Pilot Windshear Guide*. Although, primarily for the airlines, much of the information can be applied to general aviation. The Department of Commerce publication *Microbursts: A Handbook for Visual Identification* is for sale by the Superintendent of Documents. It contains an in-depth, technical explanation of the phenomena along with numerous color photographs depicting microburst activity. Avoidance is the best defense against a microburst encounter.

Weather Radar

Although an oversimplification, radar displays an image that depends on reflected energy, or back scatter.

Intensity depends on several factors; among them are particle or droplet size, shape, composition, and quantity. Next generation radar (NEXRAD) radars are capable of displaying both precipitation and cloud size particles. Note that radar detects precipitation, not turbulence.

Pilots have access to NWS radar and airborne weather radar and, to a limited extent, ATC radar. The FAA is currently providing air traffic controllers with either separate or overlay NEXRAD products. Each system has a specific purpose and its own application and limitations.

NWS radars are ideal for detecting precipitation-size particles. NWS radars can detect targets up to 250 nautical miles; however, due to range and beam resolution, which is the ability of the radar to distinguish individual targets at different ranges and azimuth, an effective range of 125 nautical miles is used.

NEXRAD has been a quantum leap in providing early warning of severe weather. Because NEXRAD is a Doppler radar, it detects the relative velocity of precipitation within a storm. It has increased the accuracy of severe thunderstorm and tornado warnings and has the capability of detecting wind shear.

Airborne weather radars have low power, generally with a wave length of 3 centimeters. Precipitation attenuation, which is directly related to wave length and power, can be a significant factor. Precipitation attenuation results from radar energy being absorbed and scattered by close targets, and the display becomes unreliable in close proximity to heavy rain or hail. Intensity might be greater than displayed, with distant targets obscured. An accumulation of ice on the aircraft's radome causes additional distortion.

CASE STUDY According to the NTSB, precipitation attenuation was a contributing factor in the crashes of a Southern Airways DC-9 in 1977 and an Air Wisconsin

Metroliner in 1980. Precipitation attenuation is not significant with NWS 10 centimeter high-power units; however, it can be a serious problem with units of 5 centimeters or less, especially in heavy rain. The NTSB recommends: ". . . in the terminal area, comparison of ground returns to weather echoes is a useful technique to identify when attenuation is occurring. Tilt the antenna down and observe ground returns around the radar echo. With very heavy intervening rain, ground returns behind the echo will not be present. This area lacking ground returns is referred to as a shadow and may indicate a larger area of precipitation than is shown on the indicator. Areas of shadowing should be avoided."

When using an airborne weather radar it is imperative to understand the particular unit, its operational characteristics, and limitations. According to the March–April 1987 *FAA Aviation News,*

> Just reading through the brochure that comes with the equipment is certainly not enough to prepare a pilot to translate the complex symbology presented on the [airborne] scope into reliable data. A training course with appropriate instructors and simulators is strongly recommended.

CASE STUDY As reported: I was the first officer and had just awoken from my rest break in first class when we hit severe turbulence. I had my seat belt on and was forced violently upward into it. Everything in front of me went up into the air, scattering pillows and other service items throughout first class. Duration was maybe 10 seconds. When I got up to the cockpit, they were trying to clean up because water had spilled on some personal effects.

We contacted dispatch and started trying to determine the extent of damage to passengers and flight attendants. There was only one episode, and there had been virtually no turbulence prior, nor was there any significant turbulence afterwards. Because I wasn't in the cockpit, I was unaware of what conditions we were flying in.

It was determined that the airplane flew into a thunderstorm during an overwater night operation that resulted in 10 seconds of severe turbulence. There was no "painting" of precipitation on radar, but lightning was observed along with St. Elmo's fire on the windshield. Because radar detects precipitation, not turbulence, and thunderstorm-related turbulence can extend up to 20 miles from the storm, pilots should be prepared for a significant turbulence encounter in the vicinity of any convective activity. The pilot's observation of lightning was a clue to the threat of turbulence.

CASE STUDY As reported: We were descending from FL290 to 15,000 ft with the autopilot on. We were deviating around a large cell at 10 o'clock. There were lower cumulus below us with tops 12,000 to 15,000 ft. We were flying south of track to avoid a large cell and miss smaller buildups. With the landing lights on we saw clouds ahead and got a short period of rain and light turbulence and one large jolt. We turned further south and broke into the clear. Four people were taken to area hospitals to be checked and released due to the one severe jolt.

The reporter went on to say, "Our radar showed nothing along our route of flight. After the encounter the controller asked how our ride was, as she had some indication of bad weather on our route." Controllers typically ask for flight conditions. The fact that a "ride report" was requested does not necessarily mean the controller was aware of hazardous weather along the aircraft's route.

CASE STUDY As reported: While cruising at FL330, the aircraft encountered severe turbulence and lost several hundred feet of altitude. There were injuries to two flight attendants and one passenger.

We were deviating around weather on a route suggested by ATC and our own flight dispatcher. We also agreed with their suggestions because we had a good picture on our airborne radar. Approximately 10 minutes prior to the encounter, we had been visual, but at the actual time of the encounter we were IFR in cirrus-type clouds. At the time of the encounter we were approximately 25 miles from the nearest contouring cell as depicted on our radar.

Because we were in clouds and radar showed us to be on a clear path, we can only assume we encountered a wind shear situation or flew into a rapidly developing buildup that did not contain enough moisture to give a return on our radar.

During a callback conversation the reporter restated the fact that the turbulence was totally unexpected. The pilot stated that the color radar that was on board did not paint the smaller cells and that might have had a bearing on the incident.

CASE STUDY As reported: Rain showers were in the area. I received a heading from ATC for weather avoidance and was advised that a previous aircraft had flown the prescribed route without incident. We had no excessive radar return and no contouring on the radar scope. Flight conditions were IFR, light rain, intermittent light chop. About 70 miles from destination we began experiencing excessive banking and pitching, which I reported to ATC as severe turbulence. I requested immediate deviation and descent because the aircraft was barely controllable. This was approved. The intensity of the turbulence remained severe intermittently for a period of about 7 minutes. During this period we reversed direction of flight and landed at an alternate.

The reporter went on to say that this incident occurred because aircraft and center radars are unable to detect wind direction changes associated with this type of tur-

Thunderstorms 93

bulence. A possible solution would be Doppler radar. Certainly Doppler radar can detect changes in wind direction and intensity, but only if precipitation is present. Doppler radar is unable to directly detect turbulence and is of no use in clear air.

CASE STUDY As reported: Along our route we encountered numerous thunderstorms and lines of cumulonimbus. We successfully navigated through the lines of cumulonimbus. We could see cumulonimbus south of us and north of our course on radar. Our radar showed several strong echoes over the arrival fix. We informed ATC and they said they could not see them. We asked for a ride report, and ATC said they had no aircraft in that area for a while. Our radar showed a clear area south of our present position. We asked and received clearance to deviate and then pick up the rest of the arrival.

I was flying the aircraft on autopilot. I initiated a right turn. We were in the clouds and a relatively smooth flight. As the turn continued, we temporarily broke out of the clouds and saw a rapidly developing cumulonimbus cloud in front of us. There was no indication of this on radar. We had excellent returns on the radar throughout the flight to this point. When I saw the cloud, I immediately disengaged the autopilot and increased the angle of bank to avoid. We penetrated the side of the cloud and experience severe turbulence for about 4 to 5 seconds. The aircraft climbed rapidly, even through I applied full forward stick. This was a brief encounter.

The reporter concluded that this encounter was unavoidable because they were unable to detect it on radar. Pilots must remember the limitations on radar. If you fly in the vicinity of convective activity, be prepared for a possible encounter with severe turbulence.

CASE STUDY As reported: While enroute we descended from FL310 to FL220 for clouds. Deviating north, of course, we rounded the corner to proceed direct. Radar showed one more area of thunderstorms for us to navigate around. Radar and visual cues showed an approximately 30-mile gap between storms with sunshine between. As we approached this area, we again received clearance to deviate. As we passed north of the first area we turned left, the clearest direction with nothing on radar. As we finished our turn we were in and out of cirrus. At 12 o'clock, less than $1/2$ mile away was a small thunderstorm with tops estimated to FL230. With no room to turn, the captain put the ignition to override just prior to entering. The aircraft started to climb. The autopilot kicked off about the time of our one big jolt of turbulence. The captain was following on the controls and in a few seconds we popped out in smooth air 200 to 300 ft above assigned altitude.

The rule of thumb is 40 miles between severe storms. The problem is that it may be next to impossible to determine the severity of a storm, especially during the development stage. Like the general limitations on radar, if you fudge on the 40-mile rule, be prepared for a significant turbulence encounter.

CASE STUDY   As reported: While on vectors with approach control we were issued a clearance to descend from 7000 to 6000 ft. The ride was rough, moderate chop to moderate turbulence and rain, with no thunderstorm activity noted on radar. Shortly after initiating the descent, we encountered a "downdraft" that, while not violent, required higher-than-normal power settings to counter. We reported "unable to hold 6000 ft" to ATC and subsequently leveled off at 5500 ft. Approach control acknowledged and advised us to return to 6000 ft when able. We regained the altitude in about a minute.

I believe we may have passed through a squall line. However, the severity was not indicated by radar presentation, PIREP, or ATC.

The reporter states they may have passed through a squall line. However, the severity of the encounter does not appear to reflect a line of severe convective activity. It does reinforce the limitations of airborne and ATC radars and the need for additional PIREPs on such activity.

CASE STUDY MSY UUA /OV NEW 150020/TM 2015/FLDURD/TP C550/TB SEV/RM occurred in area where ACFT radar did not indicate PCPN. Both crew injured.

This incident occurred over New Orleans in thunderstorm weather. Both crew of a Cessna Citation were injured when the aircraft encountered severe turbulence in an area where their airborne weather radar indicated no precipitation.

Lightning Detection Systems

Lightning detection equipment, tradenamed Stormscope, was invented in the mid-1970s by Paul A. Ryan as a low-cost alternative to radar. Stormscope and similar lightning detectors sense and display electrical discharges in approximate range and azimuth to the aircraft. Like radar, Stormscope also has limitations. One misconception proclaims that in the absence of dots or lighted bands there are no thunderstorms. However, NASA's tests of the Stormscope differed. Precipitation intensity levels of 3 and occasionally 4 (heavy to very heavy) would be indicated on radar without activating the lightning detection system. A clear display only indicates the absence of electrical discharges. This does not necessarily mean convective activity and associated thunderstorm hazards are not present. Even tornadic storms have been found that produced very little lightning. The lack of electrical activity, as with the absence of a precipitation display on radar, does not necessarily translate into a smooth ride.

Many authorities agree that a combination of radar and Stormscope is the best thunderstorm detection system. It cannot be overemphasized that these are avoidance, not penetration, devices. Thunderstorms imply severe or greater turbulence, and neither radar nor lightning detection systems, at the present, directly detect turbulence.

CASE STUDY As reported: I was flying level with my Stormscope working, talking to approach control, when without warning from my Stormscope or the controller, I experienced a sudden 800-ft increase in the altitude, very heavy rain, and heavy lightning. I at once grabbed the carb heat and the throttle. Suddenly the airplane started to descend and I lost about 1200 ft. I added power and climbed to assigned altitude and leveled off OK. The controller didn't say anything, but this was a serious altitude excursion.

We have already discussed the limitations of ATC radar. In the next section we will specifically address this issue of ATC radar.

Air Traffic Control Radar

ATC radar is specifically designed to detect aircraft, its wavelength reduces the intensity of detected precipitation. Additional features reduce the radar's effectiveness to see weather.

CASE STUDY As reported: The weather briefer gave me all available information for the flight, including thunderstorm activity in and along my original route of flight. At this point I decided to go east around the thunderstorm activity to avoid any encounters with them. I established contact with ATC for advisories and asked for radar vectors around the thunderstorm activity. They asked if I wanted to go IFR; I chose to do so, and they issued a clearance.

Within a few minutes I was in IMC and reduced to maneuvering speed. Cruise was continued with occasional light chop. Rain was encountered along with increased turbulence. I requested a lower altitude, which was denied due to "towers in my area." Cruise was continued at present altitude and heading until I saw lightning ahead of my position, also the turbulence increased in intensity. I reported this to ATC, and they asked if I wanted to turn around. I answered affirmatively and started a standard rate turn to the left to reverse my course. At this point the plane was hit by a violent updraft that registered at near 2000 ft/min. This was followed by a series of vertical shears, up and down. All gyro instruments were virtually unreadable. All control movements were for airspeed control. We were ejected from the cloud base in the fourth oscillation and in a nose down attitude.

I opted to make a precautionary landing rather than subject the aircraft, my passengers, or myself to any further danger. I felt at the time, and still do contend, that this was the only safe option. The nose gear was held off as long as possible on landing; however, the effort was stifled by a drainage ditch at the edge of the field. Considering the alternative, it was far less serious than it could have been, as there were no injuries.

The ARSR report went on the say that the reporter knew there was a line of thunderstorms and felt he could go around it. He accepted an IFR clearance because he was too close to a chance of getting into Class B airspace. When ejected from the cloud he was only a few hundred feet above the ground. Because he had carefully reviewed the route, he knew there were power lines, hills, and towers in the area.

This pilot's first assumption was that he could circumnavigate the thunderstorm activity. Without electronic storm detection equipment a pilot's only option is the "mark one eyeball" to avoid convective activity. Why did the pilot accept an IFR clearance? This eliminated the pilot's only means of storm detection and avoidance. The pilot expected ATC to vector the aircraft around the thunderstorm activity. ATC is a resource and may be able to assist the pilot around hazardous weather, but pilots can never expect the controller to keep them completely free of adverse weather or assume pilot-in-command responsibility.

The reporter states: "I opted to make a precautionary landing rather than subject the aircraft, my passengers, or myself to any further danger." Even beginning the flight appears unadvisable. When you don't have the equipment to handle the weather, don't go.

CASE STUDY As reported: On an IFR flight I was switched from approach to center after being cleared to deviate west of course. I could hear center talking to other aircraft, but they failed to respond to my check-in for some 6 to 8 minutes. During this time I entered heavy weather that did not show on the passive (lightning detection) weather indicating system. Weather included heavy rain and severe turbulence. After autopilot disconnect and slowing of airspeed, I was unable to hold altitude. Center finally responded to my calls. They told me they could not help me and switched me to approach control. After about a minute with approach, I broke into the clear.

I realize not all center radar has weather graphic overlay. If center had been able to see the weather in my flight path and if they had not been so overloaded that they could not talk to me, I would have had a safer, more comfortable flight. I also realize the limitations of the passive weather indicating system equipment.

We've mentioned the limitation of lightning detection systems and ATC radar. ATC's primary responsibility is the separation of known aircraft and the expeditious flow of traffic. Again, pilots can never expect ATC to keep them out of convective activity.

Aviation Forecasts

Forecasters can predict, within 1 to 2 hours, the onset of thunderstorms, with radar available. However, forecasters cannot predict with an accuracy that satisfies operational requirements the onset of a thunderstorm that has not yet formed. This is perfectly illustrated by most CWAs for convective activity. Some tout this product as a "nowcast." In my experience it is a "hindcast." I have yet to see one issued before the convective activity exists. This further reflects the difficulty of predicting this awesome aviation hazard.

CASE STUDY The synopsis described a moist unstable air mass. Thunderstorms were not forecast for the time of flight but were expected to develop; thunderstorms, however, were already being reported along the route. The pilot, without storm detection equipment, encountered extreme turbulence inadvertently entering a cell. The pilot, with three passengers, filed an IFR flight plan based on the fact that there were no advisories. After the encounter the pilot could not understand why a precaution or advisory regarding that system was not provided. There were no advisories in effect because, at the time of the briefing, none were warranted.

The pilot had the clues—moist unstable air and thunderstorms already reported—but put complete trust in a forecast that included no precautions or advisories.

Avoidance is the operative word with thunderstorms, microbursts, and wind shear. A pilot's proper application of many resources—training, experience, visual references, cockpit instruments, weather reports, and forecasts—makes avoidance possible.

7

Weather Systems

In general usage *weather system* is a generic term used to describe any and all weather phenomena. For our purposes a weather system is a synoptic scale event; that is, a large-scale weather pattern the size of the migratory high- and low-pressure systems of the lower troposphere with wavelengths on the order of 100 miles. The large-scale nature of weather systems has a significant effect on aviation operations. A front or frontal system is one such event, but not the only one. Upper-level lows and troughs play a key role in the evolution of weather and weather systems.

CASE STUDY As reported: A fast-moving cold front, with a small amount of moisture, moved through east Tennessee during the early morning hours, leaving visibility of 4 to 10 miles, with scattered clouds at 2000 ft, broken clouds at 2000 to 3000 ft, and overcast layers at 6000 to 7000 ft AGL. These conditions were reported between two airports 18 miles apart. Both airports are VFR only with no instrument approaches.

Between the two airports are two parallel mountain ridges. IFR flights to Mountain City are allowed to descend to 6700 ft for terrain clearance. Our company procedure allows for VFR flight to the airport when reported AWOS weather is VFR. The procedure is to request an IFR descent to the Tri Cities airport to an altitude of 3700 ft to get below the clouds and then, with good visibility, fly through a mountain pass over a lake to land at Mountain City–Johnson County Airport. Just prior to takeoff, the AWOS reported Mountain City 2700 ft scattered, 3200 ft broken, 4000 ft overcast, with 4-mile visibility. Elizabethton reported 1700 ft scattered, 2300 ft broken, 6000 ft overcast, with 10-mile visibility. Tri Cities airport reported 1300 ft scattered, 3500 ft broken, 8000 ft overcast, with 7-mile visibility.

I relied on AWOS reports from two airports 18 miles apart, but the actual conditions between the mountain ridges were different. The clouds between the mountains were lowering in the valley. My first officer was new to the company and had never flown to Mountain City. He had been a commuter pilot and was uncomfortable in these types of conditions.

Center cleared us to 6700 ft and told us to change to Tri Cities approach. I told the first officer to ask for radar vectors to Elizabethton so that we could get below the overcast and get in VFR conditions. We would proceed to Mountain City by way of Elizabethton. Tri Cities approach told us to make a 360° turn to the right and pick up a heading of 090° and descend to 3700 ft. During the turn, I asked the first officer again for Elizabethton AWOS update. He was having difficulty finding the correct frequency. Coming out of the turn to 090° we broke out of the clouds at about 4500 ft. We were flying toward Holston mountain, level at 3700 ft in what were VFR conditions, so I told the first officer to tell Tri Cities approach that we were in VFR conditions and could proceed to Elizabethton. Tri Cities approach cleared us to Elizabethton. I turned right and paralleled Holston mountain on my left. I noticed that the overcast was broken, with layers of scattered-to-broken cloud. I could see

Johnson City ahead. Visibility was probably 10 miles or greater. At the southern base of Holston mountain I saw Elizabethton Airport with a good 4000- to 4500-ft broken bases. I instructed the first officer to cancel IFR. We would proceed to Mountain City VFR.

Tri Cities approach told us a Cessna 172 had taken off from Elizabethton, climbing below us, heading west, and that it should be no factor. We were at 3500 ft, descending to 2500 ft. I told the first officer that if the conditions were not favorable, we would make a 180° turn and go back and land at Elizabethton. He agreed that we could climb if necessary.

As we got closer, I saw that the clouds were lower than expected, too low to safely continue. I immediately made a 180° turn and found myself in the clouds. I was too low to talk to approach and had to climb to get back on top and reestablish radio communications. I called Tri Cities approach for an IFR clearance. They would not issue the clearance because the Cessna 172 was about 5400 ft. We were at 4000 ft and climbing. We broke out of the clouds about 5000 ft. My first officer reported a visual sighting of the Cessna 172. I continued VFR at altitudes between 5500 and 6500 ft to stay out of the clouds. Tri Cities gave us a heading of 020°. I saw Elizabethton about 5 miles ahead. We flew over the airport and landed without further incident.

This pilot extrapolated between weather reporting stations. As the pilot discovered, conditions between reporting points, especially in mountainous terrain, can be considerably different. There is no mention of the pilot obtaining a standard weather briefing prior to the flight. Often air taxi and air carrier pilots rely strictly on their dispatch personnel to make the go/no-go decision. There is also no mention of the pilot checking with Flight Watch or FSS for updated reports, forecasts, and PIREPs. It does not appear the pilot used all available resources.

The pilot correctly deduced that both airports were VFR. Categorically they were reporting "marginal VFR." These facts should have alerted the pilot that conditions between reporting points could be less than that reported at either station and possibly below basic VFR. However, as the pilot stated, "company procedure allows for VFR flight to the airport when reported AWOS weather is VFR. . . ."

The pilot descended, canceled IFR, and proceeded VFR toward the destination. The flight crew decided if conditions were not favorable, they could make a 180° turn and go back and land at Elizabethton. From the report it appears the crew decided they could climb through the overcast, if necessary. This action involved a violation of regulations.

Conditions deteriorated, and although the pilot made a 180° turn, they ended up in the clouds. It appears the crew waited too long before attempting to extricate themselves from this situation. Now their predicament was compounded by being below radio coverage. Once in contact with ATC, a clearance could not be issued because of traffic. Fortunately, for this crew they were able to break out on top.

In the report the pilot concluded that the weather was too low to attempt such a flight. The pilot also stated, "It was poor judgement to cancel IFR, thus eliminating communications and radar contact," and that company has since changed its policy for such operations.

CASE STUDY As reported: Approaching JFK the tower had only reported moderate turbulence. However, on our departure out of 1000 to 2000 ft we encountered pockets of severe turbulence up through 10,000 ft. There were short periods of time when we were unable to control pitch or headings.

The weather in the New York area was horrible. Wind gusts of up to 45 knots with heavy rain were occurring. The runway kept changing because of the wind. Takeoff was fine for about 1 to 2 minutes; then all hell broke loose.

This air carrier crew departed into the teeth of a nor'easter. Like the previous case study it's difficult to know how much weather information was provided the flight crew. The report only contains a statement that the tower had only reported moderate turbulence. However, it is not at all unusual, due to the transitory nature of turbulence, for one crew to receive moderate intensity and then within a very short period for the next airplane to encounter severe turbulence. Even without weather advisories for severe turbulence the clues were there: strong surface winds and heavy rain.

The reporter indicated it was the worse turbulence in 26 years of flying. The pilot went on to say, "In my opinion, the weather situation should have prohibited taking off." If you are going to fly in these types of weather conditions, be prepared for the worst. As the reporter found out, sooner or later you're going to get your fillings knocked loose.

CASE STUDY As reported: After an enroute encounter with moderate icing and turbulence, our Metroliner was cleared for a VOR approach. Ice was obscuring some of our outside vision. This made it difficult to identify runway 11, with the blending town and airport lights. The lower the aircraft descended, the more frequently moderate turbulence and stronger up and downdrafts were experienced. Increased airspeed was used during final approach and maneuvering for landing. When the aircraft turned final approach, the wind was 60° to the runway gusting to 34 knots. Because of the reduced visibility out of the windshield, due to the residual ice that was still present, and the lack of a visual approach slope indicator system for the runway, the aircraft was low and below a normal glide path. On very short final, wind shear was encountered, causing the aircraft to descend dangerously low. The increased airspeed allowed the aircraft to recover, and the airplane landed safely.

The crew of this airplane came very close to being involved in an accident. The report shows how a series of relatively minor events can accumulate. Certainly turbulence and icing enroute substantially increased workload for the crew. The crew was cleared for a nonprecision approach to a runway that did not have a visual glide slope system. (The airport is equipped with ILS, but it appears the wind would not allow a landing on the ILS runway. Therefore, the electronic glide slope was of no use to the crew.) The crew had to contend with limited vision, due to windshield icing, and also with a crosswind component of 20 knots. Then the reporter states that on very short final, wind shear was encountered, causing the aircraft to descend dangerously low, in spite of additional airspeed carried on the approach.

This event occurred as a major weather system moved through the intermountain region. As is often the case, low ceilings and visibility were not a major factor. Factors associated with this incident were icing, turbulence, limited visibility due to ice on the windshield, strong crosswinds, wind shear, and the lack of a visual glide path to the selected runway. The reporter did recommend the addition of a visual glide path indicator for the runway used. This certainly would have assisted the crew, but the wind shear would still have been a significant factor. Pilots should carefully consider the advisability of accepting such an approach under these conditions. Recall the Mountain City incident. The reporter indicated that company policy had been changed. Pilots have a responsibility to themselves as well as their passengers. If an operation isn't safe, don't fly. For example, in this case, if an electronic or visual glide slope is not available under these weather conditions, perhaps the operations manual should prohibit an approach.

Like a number of the case studies involving air taxi and air carrier aircraft, if you fly in these conditions, this is what you can expect.

CASE STUDY As reported: I filed an IFR flight plan from Columbia, South Carolina to Portage County, Ohio. I was told in my weather briefing that there was a fast-moving cold front moving across the mountains. The front was producing turbulence and icing. I wanted to get back home, so I made the decision to go.

It was getting dark as I departed Columbia. After about 1 hour of fighting a strong headwind, I ran into severe turbulence. Not long after encountering the turbulence, I went into actual IFR and starting having problems holding altitude. The center gave me a block altitude of 7000 to 8000 ft. I told the center I wanted to land because I did not feel it was safe to continue the flight. Not long after that, I noticed my airspeed indicator was reading zero due to icing. I told the center I wanted to land immediately. At that time the controller gave me flight assistance to Bluefield, West Virginia, where I landed safely.

The pilot of this small general aviation airplane was aware of the weather but wanted to get back home. As well as turbulence and icing, the pilot had to contend with a strong headwind. After the loss of the airspeed was indicated, the pilot decided to call it a day.

This was a very lucky individual. This safety report could have very easily been an accident report. The pilot concluded the decision to take the flight was wrong. "With the weather conditions as they were, I should not have flown. My advice to other pilots is not to make this fight under these same conditions." The pilot also concluded it might be a good idea to turn on the pitot heat before running into visible moisture with temperatures near or below freezing.

This incident reflects the scope of weather systems. The pilot was faced with four hazards: low ceilings and visibility, turbulence, icing, and headwinds. Pilots must carefully look at the cumulative effects of weather hazards when making a go/no-go decision, as this pilot discovered—the hard way.

CASE STUDY As reported: I was on an IFR training flight. We were in VFR conditions assigned 2500 ft MSL under a 3000-ft overcast. As we were turning and slowing down, we flew from under the overcast to clear sky conditions. At this point the aircraft was pulled down, showing a 1500 ft/min descent on the vertical speed indicator (VSI). The occupants were violently thrown against the ceiling of the aircraft with such force that headsets and glasses were pulled from their heads. I applied immediate full power and 10° nose up pitch, which only slowed the descent. The severe turbulence persisted for approximately 30 seconds. As soon as the turbulence abated, we regained the lost altitude.

The weather for the flight was 2000 scattered 3000 overcast, visibility 10 miles, surface wind 180° at 15 knots. Winds aloft at 3000 ft were forecast 240° at 36 knots. Conditions existed for shear; however, the previous 12 hours of approaches and vectors in and out of clouds had been relatively smooth. There was no reason to suspect severe turbulence in good VFR conditions.

Callback initiated a follow-up. This revealed that the downdraft was encountered at the precise point that the cloud cover ended. A cold front was transiting the area. The reporter feels that the abrupt boundary of the cloud deck marked the position of the front. There are no terrain features at the point where turbulence was encountered that would suggest mechanical turbulence.

This case study reveals a pilot who had the clues and either was unable to correctly interpret them or disregarded them altogether. A cold front was moving through the area. The pilot stated surface winds were 180° at 15 knots; winds aloft at 3000 ft were forecast 240° at 36 knots. This resulted in a 60° change in direction and over a 15-knot increase in speed in less than 3000 ft. Conservatively, this results in a speed difference of at least 5 knots per 1000 ft. This is the criteria for moderate or greater wind shear turbulence. The pilot was transitioning from under an overcast to an area of clear skies. Another indicator of probable wind shear is when upward vertical motion produces a cloud, and downward vertical motion results in clear air. As this pilot found out, the distance between the upward and downward moving air can be very small, resulting in severe turbulence.

This pilot was flying in an area that was producing frontal turbulence, severe wind shear turbulence, and a boundary between upward and downward moving air. The pilot commented that conditions existed for shear. Could this be hindsight? This pilot had been conducting approaches in and out of clouds for the previous 12 hours in relatively smooth conditions. Was the pilot lulled into a false sense of security, even though the clues for severe turbulence were there?

CASE STUDY As reported: A call to the FSS indicated the weather had started to deteriorate north of Lexington, Kentucky, with lower ceilings and visibility developing. The briefer stated that VFR flight was not recommended. Dayton and Fort Wayne were at best marginal VFR. A front was moving in from the west. I elected to proceed. Somewhere south of the Falmouth VOR lower ceilings were encountered. I flew through some clouds to VFR on top right around 6500 ft. I called ATC prior to the encounter. Dayton approach called Fort Wayne to check for possible holes or VFR airports in the area. None were VFR. I had approximately 2 hours and 30 minutes of fuel left. There was still time to head south to VFR airports, with London a little over an hour south.

Why not go back? Dayton approach advised me I'd broken FAR 91. "What do you want to do?" Dayton approach asked. I replied that I'd like to go to Versailles because it had an ADF [automatic direction finder] and I also believe that I said I could land at Dayton. Dayton approach told me they could only help if I declared an emergency. This tells me that I am on my own unless I declared an emergency. I elected to declare an emergency to obtain radar assistance and separation from other aircraft. The controller provided assistance to Dayton.

The pilot received a weather briefing that indicated that a front was moving in from the west, producing low ceilings and visibilities. The pilot was advised by the briefer that VFR flight was not recommended (VNR). The pilot elected to proceed. The issuance of VNR should not in itself cause a pilot to cancel a flight, but it does indicate a need to take a very close look at the weather situation.

The pilot encountered clouds and, rather than retreating, climbed through the clouds. After contacting ATC, approach control checked for possible VFR airports in the area. Although center, approach control, and tower personnel can be helpful with a weather encounter, that is not their primary function, and they have relatively limited resources in this area. Pilots should use the FSS, preferably Flight Watch, to assess the situation well before weather deteriorates beyond their qualifications or capability.

Then, the report indicates that the pilot and ATC got into a "spitting" contest. The pilot states that ATC refused help unless the pilot declared an emergency. The pilot then states, "I elected to declare an emergency to obtain radar assistance." At this point the controller provided assistance to Dayton. In fact, an emergency can be declared by any of three entities: the pilot, air traffic control, or the person responsible for the operation of the aircraft. Although it is difficult to determine the exact circumstances out of context and with the limited information available, the pilot should have declared an emergency when it became evident that less-than-VFR conditions would be encountered. The controller should have handled this situation as an emergency, whether it was declared by the pilot or not.

Safely on the ground, the pilot spoke with ATC on the phone. The pilot states, "When the emergency was declared, the watch supervisor took over—I didn't have much choice." The pilot further says the watch supervisor said that the pilot had done a good job and didn't get rattled as most VFR pilots would have in similar situations. As a former FSS supervisor, active Aviation Safety Counselor, and flight instructor, this is not the behavior we want to compliment or encourage. In my opinion this was an inappropriate comment by the controller.

The pilot further comments that the decision to continue was incorrect and that even to start the flight in the first place was a mistake. So, how do we train pilots not to exceed their ability or that of their aircraft and to discontinue a flight when the weather goes sour? We'll discuss this aspect in the summary chapter.

CASE STUDY As reported: I was going to fly from Carson City, Nevada, to Laramie, Wyoming. I had made the trip many times and felt confident that I could make it in nearly any situation. I called FSS the night before to check weather and was told that the front that had passed through Carson that day would present some problems with mountain obscuration, low ceiling, and a freezing level just above the MEA. They also told me to expect icing in clouds. I was told that due to a strong, dry air flow, the freezing levels would be getting higher and that the system was breaking up and getting warmer as it progressed eastward.

I called FSS the next morning to file an IFR flight plan and got a weather update. The route of flight was to be

VFR from Carson to Hazen, then IFR to Ogden, Utah. Icing in clouds, turbulence, and light headwinds were forecast. The forecast freezing level was 12,000 to 13,000 ft. I figured that I could go and turn back if it got too bad. My experience with FSS weather is that it is pretty pessimistic, and I could spend the night in Ogden if the weather did not improve. I filed an altitude of 11,000 ft, the MEA. I figured that if I was below the freezing level, I would not be in the forecast area of icing and thereby OK by the FARs.

I was intermittently in IMC east of Battle Mountain, temperature of 36°F (2°C). Further east I encountered clear ice that caused a temporary loss of pitot instrument but no degradation of flight performance. I told center of my problem and asked for a vector to Elko, Nevada. Once out of the icing, I thought it would be better to go back to Battle Mountain.

The pilot obtained an outlook briefing that indicated a front had passed through the area. The FSS told the pilot of icing. The pilot's next statement is confusing. "I was told that due to a strong, dry air flow, the freezing levels would be getting higher and that the system was breaking up and getting warmer as it progressed eastward." Did the briefer mean freezing levels all along the route would be getting higher or just for the first part of the flight? Certainly a front moving eastward overnight is not going to warm significantly. Or, did the pilot misunderstand the briefer? In any case, icing remained a factor. (This emphasizes the need for pilots to have a good basic knowledge of weather theory and phenomena. A pilot cannot rely solely on the briefer. Although, some are excellent translators and interpreters of the weather, others are not.)

The pilot called FSS the next morning, filed a flight plan, and obtained a weather update. Did the pilot mean an abbreviated briefing? The pilot, by regulation, should have obtained a standard briefing for the route. The pilot stated the freezing level was 12,000 to 13,000 ft. This is not substantially different from the report the night before. Additionally, the freezing level would be lowering as the pilot approached the front.

The pilot complained about experiences of pessimistic forecasts from FSS briefers. Ironically, pilots seem to complain equally about both optimistic and pessimistic briefings. This reemphasizes the need for pilots to have a sound understanding of the weather.

The pilot planned to fly at the MEA, 11,000 ft, only 1000 ft below the forecast freezing level. In this case it seems the forecast was correct. East of Battle Mountain the pilot reports the temperature at 36°F (2°C). Unfortunately, most light aircraft temperature gauges are lucky to be within 5° of the actual temperature. From the pilot's description of the icing encounter, it appears to have only been a trace of icing, but it did cause the loss of the pitot instruments. The pilot then retreated and landed without further incident. The pilot further stated: "My familiarity with the route of flight, confidence in my abilities, and disregarding an FAR were causal."

As illustrated by all of these case studies, there is no doubt that the pressure to keep schedules for air taxi and air carrier operations, like get-home-itis for the private operation, is extremely strong. The question then becomes, Where do you draw the line? In the summary we will touch on risk assessment and management and personal minimums.

8

Aircraft Performance

Although temperature and pressure are important atmospheric properties, to understand aircraft performance a pilot must have a thorough knowledge of density. Density is the weight of air per unit volume, often expressed as pounds per cubic foot or grams per cubic meter. Occasionally, we hear the question, Which weighs less, a pound of dry air or a pound of moist air? This is the same type question as, Which weights more, a pound of lead or a pound of feathers? The answer to both is that they weigh the same—a pound is a pound. However, the volume of a pound of dry air compared to a pound of moist air, like a pound of lead compared to a pound of feathers is different. We're not really interested in these units. But, the fact that atmospheric density varies is vitally important.

As we fly higher in the atmosphere, density decreases. As noted, density is weight, and we know the higher we go the fewer molecules above; therefore, the air is less dense.

Atmospheric pressure affects density. When temperature and humidity—moisture—remain constant, if pressure is higher than standard, density is higher; conversely, if pressure is lower than standard, density is lower than standard.

Although pressure is an important factor in the density equation, temperature is often the most significant factor. Typically, the higher the temperature, the lower the density of the air. Remembering that density is the weight of molecules, at higher temperatures there are fewer molecules per unit volume, pressure and moisture remaining constant.

The final factor in air density is moisture. The higher the moisture content of the air, the lower its density because water molecules weigh less than air molecules. This is not a major factor, however. What we need to remember is that when it's humid, air density is less than when conditions are dry.

The bottom line is as follows:

- Density has a direct relationship to pressure—as pressure decreases, density decreases.
- Density has an inverse relationship to temperature—as temperature increases, density decreases.
- Density has an inverse relationship to moisture (water vapor)—as moisture increases, density decreases.

Air density affects the altimeter, airspeed indicator, and vertical speed indicator, as well as aircraft performance—including powerplant output. A pilot, to safely and effectively use the aircraft, must understand how the atmosphere affects these instruments and the aircraft.

Ram air pressure is connected to the airspeed indicator. The airspeed indicator measures the differential pressure between the ram air and static air pressure from the static ports. The altimeter and vertical speed indicators are also vented to ambient, or static, air pressure through the static ports.

Should the static ports become blocked, for example, iced over, many aircraft are equipped with an alternate static source. In an emergency the alternate static source vents the static line to the cabin. If the alternate source is vented inside the aircraft, where static pressure is usually lower than outside static pressure, selection of the alternate source may result in the following instrument indications:

- Airspeed reads greater than normal.
- Altimeter reads higher than normal.
- Vertical speed indicates a momentary climb.

The altimeter measures the altitude of the aircraft. Similar to an aneroid barometer, the altimeter measures changes in pressure as the aircraft climbs or descends. As pressure decreases, the wafers expand; through mechanical linkage the change in altitude is reflected on the altimeter face. Conversely, when pressure increases, the wafers are compressed, and the altimeter indicates a descent. If it doesn't have a means of adjustment, the altimeter indicates correct altitude only under standard atmospheric conditions of pressure and temperature.

Because atmospheric conditions vary continuously, the altimeter must be adjustable for a nonstandard environment. Most altimeters are equipped with an altimeter setting window, sometimes known as the "Kollsman window." (Paul Kollsman, a German-born aeronautical engineer, invented the method to correct the altimeter for nonstandard pressure in 1928. This was a major step in allowing pilots to fly solely by instruments.) The altimeter setting window allows the pilot to adjust the instrument for nonstandard pressure using the altimeter set knob.

The altimeter setting is a value determined for a point 10 ft above an airport (approximate cockpit height) that will correct an aircraft's altimeter to read airport elevation. It corrects the altimeter for both nonstandard pressure and temperature at the surface.

If a pilot fails to adjust the altimeter enroute and the pilot flies into an area of lower pressure, the aircraft's true altitude will be lower than the indicated altitude. Vice versa, should the pilot fly into an area of higher pressure, the aircraft's true altitude will be higher than the indicated altitude.

A final thought on altimeter settings:

CASE STUDY We were flying the Bonanza from Bakersfield, California, to Las Vegas, Nevada. A weather advisory was in effect for severe turbulence, and there were strong, gusty surface winds in the Las Vegas area. For some reason the ATIS at Las Vegas was unavailable. We contacted approach for clearance into Class B airspace and reported our altitude as 8500 ft. Approach reported our Mode C (altitude reporting transponder) readout as 8000 feet! The aircraft's Mode C transponder reports pressure altitude. The ATC computer adjusts this, based on the altimeter setting, to display indicated altitude to the controller. We had not updated our altimeter setting since leaving Bakersfield. There was a 0.50-inch difference at Las Vegas. The steep pressure gradient and lower pressure at Las Vegas were causing the strong winds and turbulence.

Nonstandard temperature affects the altimeter, although not to the same degree as pressure. Because warm air is less dense than cold air, the aneroid will display a lower altitude in cold than in warm air. Changing conditions of pressure and temperature have led to a sage saying when flying with an uncorrected altimeter: When flying from high to low (or hot to cold), look out below.

Like the altimeter, air density affects the airspeed indicator. The airspeed indicator is a sensitive, differential pressure gauge. The instrument indicates the speed of the airplane through the air—not necessarily over the ground. The airspeed indicator measures the difference between the impact pressure of the air in the pitot tube and the static pressure of the ambient air surrounding the aircraft. (The pitot tube is named after its eighteenth-century French hydraulic engineer inventor Henri Pitot.)

Indicated airspeed (IAS) is the direct instrument reading obtained from the indicator, uncorrected for installation and instrument error or variations in atmospheric density. Calibrated airspeed (CAS) is indicated airspeed corrected for installation and instrument error. These errors are usually greatest at low airspeeds, with flaps deployed. True airspeed (TAS) is calibrated airspeed corrected for atmospheric density—pressure and temperature. Because of the normal decrease in pressure and temperature with altitude, for a given indicated airspeed, true airspeed increases as altitude increases.

All aircraft have limitations, including turbojet aircraft that often fly at the edge of their performance envelope. Nonstandard conditions can critically affect performance. The pilot's task is to determine aircraft performance based on actual or forecast conditions.

Now let's put the "theory" into practical application. Aircraft operations can be divided into takeoff, climb, cruise, descent, and landing.

High temperatures at high-altitude airports produce high-density altitude. Atmospheric pressure and humidity are also factors; however, temperature and elevation are paramount. Surface temperature forecasts are not normally available, but maximum temperatures typically occur beginning by midmorning and continuing through late afternoon. For aircraft with limited performance, arrival and departure times must be planned based on aircraft performance.

Under certain conditions the pilot can get the airplane off the ground only to be trapped in ground effect.

Ground effect is the temporary gain in lift during flight at altitudes of about the wing span of the airplane, due to compression of the air between the wing and the ground. If this is allowed to continue beyond the end of the runway, there is only one result—an airplane accident!

The following is a typical scenario: A pilot attempts to take off from a relatively short runway, during midday, with an aircraft at or above maximum gross takeoff weight. The aircraft accelerates and lifts off and may initially establish what the pilot perceives as a positive rate of climb. However, the aircraft then begins to settle and, with no runway remaining, hits terrain. Pilots often blame downdrafts or windshear for this when weather conditions are conducive to neither.

CASE STUDY As reported: The pilot and three other pilots were planning to fly to an aviation safety seminar in a Piper Arrow. Flaps were set to 10° and full power was applied, using a rolling takeoff. The pilot rotated near the end of the 3000-ft strip. The airplane climbed to about 40 ft before settling into trees. An investigation revealed the airplane was about 175 pounds over maximum gross takeoff weight, density altitude was 4500 ft, and the runway had a 0.5 percent upslope grade. The *Pilot's Operating Handbook* (*POH*) recommends a 25° flap setting, with full power before break release for an obstacle clearance takeoff.

The NTSB determined the probable cause as the pilot's inadequate planning, preparation, and takeoff technique.

There were about seven factors that contributed to this incident. Planning—the pilot apparently did not calculate takeoff requirements as required by the regulations, the airplane was over gross weight, and the runway was upslope. Preparation—the pilot failed to calculate density altitude and did not select an abort point based on these calculations. Takeoff technique—the pilot failed to select the proper flap setting and did not use the manufacturer's recommended obstacle clearance technique. The elimination of any one of these errors might have prevented this very embarrassing accident. As we'll see in the next chapter, a pilot's fate often rests in his or her own hands.

Most flight computers provide for density altitude calculations. There are also numerous graphs and charts that provide this information.

CASE STUDY I have flown Cessna 150s out of Bryce Canyon, Utah (elevation 7586), and Grand Canyon, Arizona (elevation 6606), and a Turbo Cessna 210 out of Mammoth Lakes, California (elevation 7128 and density altitude 10,000 feet). There is no additional hazard in such operations as long as we calculate, and do not attempt to exceed, aircraft performance.

After calculating that the aircraft has sufficient performance for conditions—which we all know is required by regulations—it's a good idea to determine an abort point on the runway. If the aircraft is not airborne, climbing out of ground effect, with a positive rate of climb, this point should allow the pilot to come to a safe stop on the remainder of the runway. Remember, aircraft performance data are based on a new airframe and engine and perfect pilot technique.

CASE STUDY The pilot of a Beech A36 Bonanza attempted to depart Las Vegas, New Mexico, from a 5000-ft strip, density altitude 8800 ft. After two aborted takeoffs, due to what the pilot reported as engine problems, a third try was attempted. The pilot tried to get the airplane airborne at the end of the runway but was unsuccessful. The airplane hit the ground and rolled off the runway. The airplane *POH* indicated the takeoff distance should have been about 2000 ft.

This is an example of a pilot's failure to determine the required takeoff distance and then abort the takeoff at midfield if sufficient speed was not attained.

Normally, a pilot of a carbureted, fixed-pitch propeller airplane will adjust the mixture, prior to takeoff, for maximum engine rpm. Some pilots run the engine to maximum power during the runup and then adjust the mixture. (Cessna recommends this procedure when density altitude reaches 3000 ft. However, always check the *Pilot Operating Handbook* for the airplane you're flying for recommended procedures.) I prefer to adjust the mixture to a rough setting at runup rpm and then during the initial takeoff roll make final mixture adjustments. This procedure puts extra workload on the pilot during this critical phase but less stress on the airplane.

A pilot of a nonturbocharged, constant-speed propeller airplane will use the procedure described above, except lean for maximum manifold pressure. The pilot must realize that manifold pressure will be reduced proportionally to density altitude. For example, with a density altitude of 6000 ft, the pilot could only expect 24 inches of manifold pressure at full throttle.

An advantage of a turbocharged engine is the development of sea-level power at altitude. The turbo Cessna 182 *POH* recommends leaning to smooth engine operation.

On airplanes equipped with a fuel flow indicator, for example, the Bonanza, the pilot can lean to density altitude takeoff fuel flow. This setting is a rough estimate, and smooth engine operation should always be maintained.

Multiengine pilots must consider that airport density altitude may exceed inoperative engine ceiling. For example, an early model nonturbocharged Cessna 310 has a single-engine ceiling of approximately 6800 ft; a Piper Aztec, 7500 ft; and a Piper Seminole, only 5000 ft. Remember, these altitudes apply to standard conditions. Density altitude in the summer, especially in the west, can often exceed these values.

Pilots cannot allow themselves to be lured into a false sense of security because the airport is at a relatively low elevation.

CASE STUDY The pilot of a Piper Saratoga, a turbocharged airplane, elected to depart from a 1913-ft grass strip in Texas, field elevation less than 1000 ft, air temperature 34°C. The density altitude was approximately 3000 ft. The grass was wet and from 4 to 6 inches tall. The aircraft struck power lines that were about 30 ft above the ground. The aircraft was slightly over gross weight. Based on the *POH*, with a dry runway, the aircraft required 1700 ft to clear a 50-foot obstacle. For a wet, sod runway the distance required would be approximately 2300 ft!

Most density altitude accidents involve improper takeoff procedures. In nearly half of these accidents the pilot attempted to take off in excess of maximum gross weight. In about one-third of them the pilot attempted to exceed the climb performance of the aircraft. All were preventable.

Climb and cruise performance are also affected by density altitude.

CASE STUDIES We had filed an IFR flight plan from Lancaster's Fox Field in California's Mojave Desert to Ontario, California. I requested 7000 ft and planned to go through the San Fernando Valley because of lower minimum altitudes. The clearance came back, "cleared via the Cajon two arrival; climb and maintain 11,000." The surface temperature was 86°F (30°C) and the Cessna 150 was not going to 11,000 that day. After negotiating with a rather perturbed ground controller, I received my requested routing. Pilots must know their aircraft's performance and not allow ATC, or anyone else for that matter, to push them into an untenable—in this case, unobtainable—position.

Let's take another example, on this particular day the average temperature from Oakland to South Lake Tahoe at 13,500 ft was 25°F (–04°C). Standard temperature for 13,500 ft is 14°F (–10°C). The forecast temperature was 6° warmer, which is above standard. When air is less dense than standard, aircraft performance will be less than performance charts for standard conditions advertise. Based on the above conditions, density altitude is about 15,000 feet.

CASE STUDY My 1966 Cessna 150 had a book ceiling of 12,650 ft. We planned to traverse the 9943-ft Tioga Pass in California's Sierra Nevada range. The winds were out of the northeast at only 10 knots, resulting in a slight downdraft from the wind flowing up the east slopes and down the west slopes as we approached the pass. Temperature was slightly above standard, and in combination with the wind, the airplane wouldn't climb out of 9500 ft. We had to proceed north along the west slopes of the mountains to Ebbett's Pass at 8732 ft, where we were able to safely cross the mountains.

Gain the required altitude prior to reaching the pass or crest of the mountains, with sufficient room to make a comfortable course reversal, if required. One technique is to approach the ridge line at a 45° angle. If the ridge cannot be cleared, the pilot usually only has to make a 45° turn away from the mountains to reach lower terrain. How can we tell if we'll clear the crest? If we're above the crest, the terrain beyond will appear to be descending in relation to the crest of the mountains.

An advertised service ceiling of 13,100 ft for an aircraft is based on standard conditions. Differences are usually not significant unless the pilot is operating at the limit of the aircraft's performance. Unfortunately, this occurs every year with pilots who attempt to cross the Sierra Nevada or Rocky mountains in conditions well above standard. Some pilots can't understand why an aircraft with a service ceiling 13,100 ft can't climb above 12,000 ft with an outside air temperature of 32°F (0°C). Density altitude at 12,000 ft and temperature 32°F (0°C) is 13,100 ft. And, this is an ideal case. A runout engine, poor leaning technique, over gross weight, and the possibility of turbulence and downdrafts would further decrease performance. Some have mused that you can walk across the Rockies and Sierra Nevada mountains on the wreckage of Cessna 172s and Piper Cherokees.

Descents are normally not a problem. But, don't forget to richen the mixture, if required.

A pilot should use the same indicated airspeed values for a high-density altitude approach and landing as those used at sea level. because instrument indications are based on air density. But, airplane speed over the ground (ground speed) will be higher. This is one reason for longer ground rolls at higher-density altitudes. Multiengine pilots, especially, should know density altitude prior to landing. If a go-around should be required, the pilot should know if the airplane is above its single-engine ceiling.

The bottom line: Calculate existing density altitude and don't expect the aircraft to exceed its design limitations. Most flight computers have density altitude functions. Various tables and charts have been developed for this purpose. There is no excuse for not obtaining and applying this information.

9

Summary

Many have touted the necessity for pilots to get the "big picture." This refers to obtaining a complete weather synopsis; that is, the position and movement of weather-producing systems and those that pose a hazard to flight operations. This is important, but it is only one element needed for an informed weather decision. I prefer the term *complete picture*. As well as a thorough knowledge of the synopsis, pilots must obtain and understand all the information available from current weather to forecast weather. Each report, chart, or product provides a clue to the complete picture. Each must be translated and interpreted with a knowledge of its scope and limitations. Then, with a knowledge of the complete picture we can apply the information to a specific flight. As observed by the United States Supreme Court: "Safe does not mean risk free." With the complete picture, a knowledge of our aircraft and it's equipment, and ourselves and our passengers, we're ready to assess and manage risk.

As we've seen in previous chapters, the need to evaluate all available sources cannot be overemphasized—adjacent weather reports, PIREPs, radar reports, satellite imagery, and forecasts. The complete picture and a knowledge of weather can help with evaluation and to identify erroneous observations. Both manual and sensor errors can be detected in this way. Whether it is a manual or automated observation, pilots must remember that reported conditions may not be representative of the surrounding area.

As an old aviation axiom laments, aviation weather reports may not be accurate, but they're official. The question is not whether manual or automated reports are better or worse. They are different, with both having advantages and limitation that must be understood.

Part of the complete picture is having more than one way out. When only one out is left, it's exercised. This might mean canceling a flight, circumnavigating weather, avoiding hazardous terrain, or making an additional landing enroute. The 180° turn is made before entering clouds or hazardous weather. If the situation becomes uncertain, assistance is obtained before an incident becomes an accident. Armed with the complete picture a pilot will know the nearest area of good weather, should the forecast go sour.

A competent pilot will never be caught on top or run out of fuel. These pilots combine mental attitude and skill to update weather enroute, devise a plan based on this information, and coordinate the action before the situation becomes critical.

There are a number of advantages to self-briefing. For the weather-wise pilot, experienced in accessing and decoding the hieroglyphics of aviation weather, it's often easier to get the complete picture for oneself than to talk with an FSS controller. This is not to criticize; it's just that FSS controllers are trained to distill the weather picture for pilots, and some pilots would rather see the big picture in its entirety for themselves. Computer access to weather information gives pilots all the raw data, which is often helpful in answering tough go/no-go decisions. Another advantage is that, assuming you can log into the Internet without being "put on hold," it's often quicker to fire up the computer than it is to reach a live FSS person on the telephone. This is especially true during peak flight planning times and during bad weather. Access to color graphics is another distinct plus with most self-briefing services.

On the other hand, many FSS controllers are excellent interpreters and translators of the weather. They are a resource that should not be overlooked. They can, often, provide valuable assistance to the pilot, especially if any information is missing or unclear.

The importance of updating weather and NOTAMs enroute is another area that cannot be overemphasized. The primary focal points for these services are the FAA's Flight Service Stations, through FSS communications, broadcasts, and Flight Watch. Secondary sources of information are Automatic Terminal Information Service (ATIS) and automated weather observation (AWOS/ASOS) broadcasts, and Air Route Traffic Control Center (ARTCC), and terminal (tower/approach control) controllers.

The early days of airline flying were plagued by thunderstorms as well as icing, turbulence, widespread low ceilings and visibilities, and the limited range of the aircraft. Today's jets have virtually overcome these obstacles. More and more pilots of general aviation aircraft, equipped with turbochargers and oxygen or pressurization, are encountering the same problems as yesterday's airline captains. The only difference is a vastly improved air traffic and communication system. Among one of the

FAA's best-kept secrets has been the implementation of high-altitude Flight Watch.

Continually updating the weather picture is the key to managing a flight, especially at high altitude in aircraft without ice protection and storm detection equipment and with relatively limited range. Winds aloft can be a welcome friend eastbound or a terrible foe westbound. With limited range even a small change in winds at altitude can have a disastrous result. At the first sign of unexpected winds, Flight Watch should be consulted, if for no other reason than to provide a pilot report. A significant change in wind direction or speed is often the first sign of a forecast gone sour. A revised flight plan might be required. Flight Watch can provide needed additional information on current weather, PIREPs, and updated forecasts upon which to base an intelligent decision.

A primary reason for high-altitude flying is to avoid mechanical, frontal, and mountain wave turbulence; however, the flight levels have their own problems—wind shear or clear air turbulence. When problems are encountered, Flight Watch can help find a smooth altitude or alternate route. If the pilot elects to change altitude, an update of actual or forecast winds aloft is often a necessity.

Icing is normally not a significant factor in the flight levels, except around convective activity or in the summer when temperatures can range between 14°F and 32°F (–10° and 0°C). However, icing can be significant during descent, especially when the destination temperature is at or below freezing. Flight Watch can provide information on tops, temperatures aloft, reported and forecast icing, and current surface conditions.

Many aircraft are equipped with airborne weather radar and lightning detection equipment. However, these systems are plagued by low power, attenuation, and limited range. A pilot might pick his or her way through a convective area only to find additional activity beyond. Flight Watch has the latest NWS weather radar information. Well before engaging any convective activity, a pilot should consult Flight Watch to determine the extent of the system, and its movement, intensity, and intensity trend. Armed with this information, the pilot can determine whether to attempt to penetrate the system or select a suitable alternate. ATC prefers issuing alternate clearances compared to handling emergencies in congested airspace and severe weather.

Finally, there is destination and alternate weather. The preflight briefing provided current and forecast conditions at the time of the briefing. This information should be routinely updated enroute; the airlines do it, often through Flight Watch. Are updated reports consistent with the forecast? If not, why? Flight Watch controllers through their training are in an excellent position to detect forecast variance. Whether the forecast was incorrect or conditions are changing faster or slower than forecast, the pilot needs to know and plan accordingly. A knowledge of forecast issuance times is often helpful. Forecasts might not be amended if the next issuance time is close. Flight Watch is in the best position to provide the latest information and suggest possible alternatives.

Updates must be obtained far enough in advance to be acted upon effectively. This must be done before critical weather is encountered or fuel runs low. Hoping a stronger-than-forecast head wind will abate, or arriving over a destination that has not improved as forecast, is folly. At the first sign of unforecast conditions, Flight Watch should be consulted and, if necessary, an alternate plan developed. This might mean an additional routine landing enroute, which is eminently preferable to, at best, a terrifying flight or, at worst, an aircraft accident.

Personal minimums are directly related to risk assess and management. Aeronautical decision making is defined as the ability to obtain all available, relevant information, evaluate alternate courses of actions, then analyze and evaluate their risks, and determine the results. First, evaluate all the factors for a particular flight and decide if the risk is worth the mission. This can be extremely simple or extremely complex. There are three elements in risk assessment and management: planning, aircraft, and pilot. Planning is the "homework" part of the flight. We study terrain, altitude requirements, and the environment. The environment includes the weather, our personal minimums, and alternatives. Now we evaluate the aircraft. Does it have the performance and equipment for the mission? If the answer is Yes, we preflight the aircraft and determine that it is airworthy. Assuming the pilot is "fit for flight," we're ready to go.

There are many excuses for getting caught in weather, but few if any reasons. With our knowledge of the three-dimensional atmosphere, we should be able to put together the complete picture. Understanding why some weather systems are benign and other severe, and how they are modified, and then integrating this knowledge with the preflight weather briefing and updates enroute will allow us to make intelligent, safe weather decisions. But, this is only half of the safety equation. We have reviewed numerous weather-related incidents and accident scenarios. Most frequently an incident or accident occurred because the pilot failed to obtain complete and accurate information, attempted to exceed aircraft performance, or simply continued flight beyond its capability or below safe minimums. Virtually all were preventable.

Summary 139

We know about certification and operating rules. We have a complete picture of the weather. We know our aircraft and it's equipment and ourselves and our passengers. How do we decide if a particular flight is safe? How do we assess the risk and manage that risk? We've discussed various scenarios thus far.

Unlike being a "little pregnant," there is some middle ground when it comes to the weather. For example, we can plan the flight in stages, landing short of our ultimate destination. Then we can take another look at the weather. However, there are two caveats to this option. First, we must know when to abandon the plan. When it's not meant to be, it's not meant to be. We must know when to call it a day. Second, we must have an alternate plan or two (plan B, plan C, plan^n).

A new pilot certificate or rating should be considered a "ticket to learn." Student pilots are guided through their initial training under the direct supervision of a flight instructor. Military pilots are shepherded by more experienced pilots. And the airlines have an extensive dispatch system to ensure the safety of their flights. However, after initial certification, there is no such safety umbrella for the general aviation pilot. This fact is directly reflected in the accident record. In a very real sense we hold our fate in our hands.

To apply a weather briefing to a flight we must have done our homework. What is the terrain like along the route? What are the minimum altitudes? Are there suitable alternates? What if plan A does not pan out? It's incumbent upon the pilot—for every briefing—to study the terrain, routes, and possible alternates for the proposed flight. For example, recall my experience with Los Angeles Center. I had planned the flight below the freezing level, but due to traffic I had to climb into icing conditions. The objective is to have an out. If there are no outs, the flight is a definite no-go.

Accident prevention is part of NASA's commitment to aeronautics. To this end they have developed scenarios of precursors to aviation accidents. A *precursor* is a factor that precedes and indicates or suggests that an incident or accident will occur. It might be physical incapacity, poor judgment, aircraft deficiency, failure of the ATC system, the weather, or other factors that by themselves would not create an incident or accident, but when taken together, they lead to disaster.

CASE STUDY Seven-year-old Jessica Dubroff accompanied her father (a passenger) and the pilot in command in an attempt for a so-called transcontinental record involving 6660 miles of flying in eight consecutive days. (I say so-called record because this was nothing more than a publicity stunt. It reminds me of telling friends that my son soloed at the age of 3 months. He was the sole occupant of the airplane as we pulled it over to the wash rack.) The first leg of the trip, about 8 hours of flying, had been completed the previous day, which began and ended with considerable media attention.

On the second day they participated in media interviews, preflight, and then loaded the airplane. The pilot in command received a weather briefing that included weather advisories for icing, turbulence, and IFR conditions, due to a cold front moving through the area.

The airplane was taxied in rain for takeoff. While it was taxiing, the pilot acknowledged receiving information that the wind was from 280° at 20 gusting to 30 knots. A departing Cessna 414 pilot reported moderate low-level wind shear of ±15 knots. The airplane departed toward a nearby thunderstorm and began a gradual turn to an easterly heading.

Witnesses described the airplane's climb rate and speed as slow, and they observed the airplane enter a roll and descent that was consistent with a stall. Density altitude at the airport was 6670 ft. The airplane's gross weight was calculated to be 84 pounds over the maximum limit at the time of impact.

Summary

The probable cause was the pilot's improper decision to take off into deteriorating weather conditions. This included turbulence, gusty winds, an advancing thunderstorm, and possible carburetor and structural icing. The airplane was over gross weight. Density altitude was higher than the pilot was accustomed to. The result was a stall caused by failure of the pilot to maintain airspeed.

As in virtually all the previous examples, most incidents and accidents can be attributed to a series of relatively insignificant factors that when taken together cause the situation. Let's review the Dubroff accident in this context.

They were on a tight schedule. Publicity events had been scheduled in advance. The original takeoff time was delayed to allow Jessica additional sleep. The pilot was fatigued from the previous day's flight and obtained little rest during the night. The weather was marginal at best. The pilot had to obtain a special VFR clearance for departure. Who was really flying the airplane? The pilot in command was seated in the right seat of the Cessna Cardinal. Now add high-density altitude, an overloaded airplane, and a mind-set that they must go.

The first precursor was the need to keep a time schedule—sometimes referred to as "get-home-itis." Precursor number 2 was pilot fatigue. The next precursor was a high-density altitude takeoff with an airplane over gross weight. The fourth precursor was the weather, with its low ceilings and visibility, gusty winds, wind shear, turbulence, icing, and thunderstorms. (You could count each of these weather factors as an individual precursor.) A fifth precursor was the pilot's attempt, under these very adverse conditions, to try to maintain control of the airplane. We will never know what exactly happened, but airplane control was lost. The deck was certainly stacked against them.

Like most accidents, I think we can see how breaking any one individual link could have prevented this accident. The first link was the time schedule. A friend, an excellent pilot, has the philosophy that there is never a reason that you absolutely have to be anywhere. In the Dubroff case the pilot's mind-set appeared to be: We're going, no matter what.

The second link was fatigue. It was reported that Jessica had slept most of the first leg. Pilot fatigue is a significant factor in the deterioration of both mental and physical skills. This certainly may have clouded the pilot's go/no-go decision and apparent failure to calculate gross weight and density altitude.

The weather was terrible. If the weather had been clear and calm, the pilot might have gotten away with fatigue, overloading the airplane, and lack of experience with high-density altitude.

Now add the pressure of flying from the right seat, with a novice student in the left, in less-than-basic VFR conditions. Even a slight, momentary distraction under these conditions can have serious consequences. It's reasonable to conclude that the pilot experienced sensory overload during climb out. All of these factors together aligned the precursors, resulting in a fatal accident.

So how do we assess and manage risk? We apply aeronautical decision making. Our goal is to prevent the precursors from aligning. That's easy for me to say. The decision can be as simple as my friend John and his Kit Fox looking at an afternoon flight in the traffic pattern or as complex as one of NASA's shuttle missions.

Let's start with John's decision. Planning: Airport elevation 397 feet, runway 25L 2699 ft; pattern altitude 1400 ft; the environment—clear, cool, winds calm, alternate runway 25R. Aircraft: Performance of the Kit Fox OK; airplane equipped for flight in Class D airspace. Pilot: Fit for flight. Decision: go.

Don't worry. We're not going to evaluate a space shuttle mission. Instead let's take an actual flight. We were flying from Oklahoma City to Palm Springs for the 1998 AOPA convention. The weather was good through Tucumcari, New Mexico, but deteriorated between Tucumcari and Albuquerque. Passing Tucumcari I checked with Flight Watch and received the bad news. The weather ahead was IFR to marginal VFR (MVFR). The plan was to fly direct, via the Anton Chico and Otto VORs. With the deteriorating weather ahead, I decided to go IFR—I follow roads. With little in way of landmarks, low ceilings and visibilities, the safest option was to follow I-40. The terrain and clouds merged about 20 miles west of Santa Rosa, New Mexico. It was afternoon and we had been flying for about 4 hours. With night approaching, poor weather, and fatigue a factor, the only viable option was to return and land at Santa Rosa.

Risk assessment and management does not stop with a go decision. Should the airplane be unairworthy—this includes equipment—for the flight, the decision is no-go. Conditions must be reevaluated throughout the flight. If the situation at the destination (wind, weather, surface conditions, etc.) changes, we may have to divert. If we don't have an alternate plan, the risk is too high, resulting in a no-go decision. What if the aircraft should become unairworthy during flight? As stated in the regulations: "The pilot in command shall discontinue the flight when unairworthy mechanical, electrical, or structural conditions occur."

CASE STUDY After remaining in Santa Rosa, New Mexico for 2 days, the weather had not significantly improved. With high minimum altitudes and a low freezing level, IFR flight was a definite no-go.

With the preflight complete, $4\frac{1}{2}$ hours of fuel, we departed Santa Rosa and opened our VFR flight plan to

Albuquerque. (A VFR flight plan, especially under these conditions, is part of risk management.) Ceilings were low, but visibility was excellent. It soon became apparent that plan A, over the clouds, was not going to work. This was confirmed through a conversation with Albuquerque radio, advising that their weather had not improved. Plan A: no-go.

Plan B—fly under the clouds. Approaching Clines Corners, terrain rises to about 7000 ft. The clouds went right down to the ground. When, as pilots, do we say "No" and call it a day? I teach, or maybe it's preach, that the first time the thought "Should I really be here?" or "Maybe I should turn around" occurs is a red flag to take immediate positive action. Don't push the weather, your aircraft, or yourself; turn around and wait it out. We initiated a 180° turn. We would have been flying from poor weather to worse weather—the risk was too high and the decision was no-go.

At this point I had resigned myself to returning to Santa Rosa—plan D. However, my wife said, "What about plan C?" An increased risk accompanied plan C. There were only a couple of dirt strips with high elevations and short runways for alternates. The terrain was lower, ceilings low, but visibility remained excellent. For navigation we had the "iron compass" (railroad). I called Albuquerque radio and changed our route and ETA. As is my practice I made position reports and updated weather with Flight Service—another part of risk management. We always had the option of returning to Santa Rosa should the weather deteriorate.

Albuquerque did not improve and we landed short at Alexander, New Mexico. With the weather now improving from the west, the flight continued uneventfully on to Palm Springs.

CASE STUDY The Cessna 172 pilot received a preflight weather briefing that included marginal VFR conditions and reported icing in clouds near the route of flight. The

Summary

pilot pulled the carburetor heat control to the ON position and descended to about 500 feet AGL to maintain visual contact with the ground. About $1/8$ inch of ice had formed on the airplane, and the pilot reversed course in an attempt to locate an airport. The flight controls felt "sluggish." The pilot selected a field, configured the airplane, and made a precautionary landing. During the landing, the nose gear sank into the muddy field, then collapsed, and the airplane nosed over.

The NTSB determined the probable cause to be the pilot's continued flight into adverse weather. They cited as factors the low ceiling, icing conditions, airframe ice, and the muddy field—an alignment of at least four precursors. Unfortunately, this is another example of a situation where the pilot failed to retreat before entering adverse weather conditions. I know it's easy for me to say, but the goal is not enter these conditions in the first place.

Throughout the previous chapters reporters have mentioned their hesitation to declare an emergency. An emergency can be a distress or an urgent situation. *Distress* is a condition of being threatened by serious and/or imminent danger, requiring immediate assistance. *Urgent* is a condition of being concerned about safety and requiring timely, but not immediate, assistance—a potential distress situation. Controllers would much prefer that pilots declare an emergency and obtain assistance before a bad situation becomes an impossible one.

CASE STUDY It was the second day of the trip back from the Oshkosh Fly-In. We had departed Gillette, Wyoming, for Pocatello, Idaho, in a Bonanza. We planned to use VOR and pilotage navigation through Jackson Hole and Idaho Falls. The weather was not a significant factor, but visibility was restricted due to smoke of numerous forest fires. In this part of the country, even at 12,500 ft we were still below the peaks. After passing what I identified as the Grand Tetons we turned southwest toward Pocatello.

Well, you guessed it, we couldn't receive any VOR or establish communications with any facility. Continuing to fly down what I thought was the Snake River Valley, things didn't seem quite right. I was following an old aviation axiom: Follow a river or a road and it will normally bring you to a town, and maybe an airport.

Even with $2\frac{1}{2}$ hours of fuel, we decided it was time to resolve the issue of position. Because we were unable to establish communications on standard frequencies, I selected 121.5. We were not in distress, but there was a sense of urgency. Therefore, as outlined in the *Aeronautical Information Manual* (*AIM*) I broadcast "PAN PAN PAN" followed by the aircraft identification. Almost immediately a military Air Evac flight responded. Based on our assumed position, the Snake River Valley, Air Evac provided us with a frequency for Salt Lake Center.

After several tries, however, we were unable to establish communication. By this time we had come across a small town with a good-sized airport. Unfortunately, there was no name on the airport. Because our transponder was being interrogated, we knew someone had us on radar. I selected 7700 and again broadcast "PAN PAN PAN." Air Evac again responded. I requested Air Evac to ask center to look for a 7700 squawk. In a few moments Air Evac responded with another center frequency.

Calling center, the controller immediately responded, "Your position is 6 miles east of Big Piney." That left us with one minor question: Where is Big Piney? After a few moments shuffling the charts we were on our way, although not by the route originally planned. As Maxwell Smart would say, "Missed it by that much!" Well, it isn't much on a world aeronautical chart (WAC).

Summary

This incident illustrates several important points. Enroute, monitor the emergency frequency when you have a second radio. Keep careful track of your position and fuel. And, should a situation of uncertainty develop, don't hesitate to request assistance. The world doesn't come to end by using 121.5 or squawking 7700. Unless there is some obvious pilot inadequacy, such as running out of fuel or flying into adverse weather, the FAA isn't going to get involved. Student pilots especially shouldn't fear requesting assistance. The FAA isn't going to talk to them; they're going to want to speak to their instructor. I know this from personal experience. For my part, my students always know when and where to obtain assistance, as should every pilot. The goal is to prevent an incident from becoming an accident.

The FAA's policy is that pilots who become apprehensive for their safety for any reason should *request assistance immediately.* Ready and willing help is available in the form of radio, radar, direction finding stations, and other aircraft. Delay has caused accidents and cost lives. *Safety is not a luxury! Take action!*

Index

Aeronautical Information Manual (AIM), 57, 146
Aircraft Owners and Pilots Association (AOPA) Air Safety Foundation, 4, 8
AIRMET ZULU, 77
Airplane Flight Manual (AFM), 79 (*See also* Pilot's Operating Handbook)
Area Forecast (AF), 6
ASOS (*See* Automated observations)
ASOS operational assessment, 20
Automated observations, 13, 17, 19, 24, 37, 104, 106, 135
Automated Lightning Detection and Reporting System (ALDARS), 17
Automatic Terminal Information Service (ATIS), 76–77
Aviation Weather (AC 00-6), 7
AWOS (*See* Automated observations)

Center Weather Advisory (CWA), 77, 99
Clear air turbulence (CAT), 49–51
Cloud height indicator (CHI), 17, 19
Cockpit resource management (CRM), 78
Code of Federal Regulations (CFR), 68, 71
Complete picture, 133
Cooperative Program for Operational Meteorology, Education and Training (COMET), 4

Direct User Access Terminal (DUAT), 67
Distress, 145 (*See also* Emergency)

Doppler radar, 93 (*See also* NEXRAD)
Dubroff, Jessica, 140–142

Emergency, Declaring an, 71–72, 74, 81, 83, 112–113, 145–147
Enroute Flight Advisory Service (EFAS), 5, 30, 105, 136–137

FAA's aviation weather policy, 3
FAA's tombstone mentality, 55
Federal aviation regulations (*See* Code of Federal Regulations)
Flight Service Station (FSS) 11, 30–31, 3867, 105, 113–115, 134–135
Flight watch (*See* Enroute Flight Advisory Service)
Freezing precipitation, 80

Garvey, Jane F., 3

Icing, known (*See* Known icing)

Kelly, John J. Jr., 4
Known icing, 83
Kollsman, Paul, 121

Low-level wind shear (LLWS), 7, 87–88
METAR, 7
Microbursts: A Handbook for Visual Identification, 88
Minimum equipment list (MEL), 79

National Aeronautics and Space Administration (NASA), 4, 78, 95, 140, 142
National Aviation Weather Program Plan, 1992, xviii

National Aviation Weather Program Strategic Plan, April 1997, xviii
National Transportation Safety Board (NTSB), 29, 31, 34, 54, 74, 87, 90, 145
NEXRAD, Next generation radar, 89
Notice to Airmen (NOTAM), 20, 67–68, 135

Packing effect, 17
Personal minimums, 138
Pilot's Operating Handbook (POH), 124, 126–127
(*See also* Airplane Flight Manual)
Pilot weather report (PIREP), 19, 34, 37, 40, 46, 74, 77, 94–95, 105, 134, 136
Pilot Windshear Guide (AC 00-54), 88
Pitot, Henri, 123

Scud run, 37
Shaw, Sir William Napier, 5
SIGMET, 48, 77
SPECI, 12
St. Elmo's fire, 91
Stormscope, 95

Sucker hole, 30
Supercooled large drop (SLD), 77

Tailplane stall, 75
Terminal aerodrome forecast (TAF), 6–7
Traffic alert and collision avoidance system (TCAS), 50, 55, 57
TWEB route forecast, 6

UNICOM, 30–31, 68
United States Supreme Court, 133
Urgent, 145–146
(*See also* Emergency)

VFR flight not recommended (VNR), 41, 112
Visibility:
  automated, 12
  prevailing, 12

Wake turbulence separation minima, 52–53, 60–61
Weight categories (*See* Wake turbulence separation minima)
Wind shear, 46, 110
  (*See also* Low-level wind shear)

About the Author

Terry T. Lankford was a Weather Specialist with the FAA for nearly 30 years, and is the author of three books in McGraw-Hill's *Practical Flying Series: Understanding Aeronautical Charts*, Second Edition, *Cockpit Weather Decisions*, and all three editions of *Weather Reports, Forecasts, and Flight Plans*. He also wrote *Aircraft Icing*. A pilot since 1967, he holds single-engine, multi-engine, and instrument ratings, as well as an FAA Gold Seal Instructor certificate. An FAA accident prevention counselor, he earned the Flight Safety Award in 1979. Lankford also contributes articles to pilot periodicals.